Pre-Contract Practice and Contract Adminis

for the Building Team

Also available
Tenders and Contracts for Building
Third Edition
The Aqua Group
0-632-04277-X

Pre-Contract Practice and Contract Administration
for the Building Team

The Aqua Group

Edited and updated by

Mark Hackett & Ian Robinson
of Davis Langdon & Everest

Blackwell
Science

© 2003 The Aqua Group and Blackwell Science
Ltd, a Blackwell Publishing Company
Editorial Offices:
Osney Mead, Oxford OX2 0EL, UK
    Tel: +44 (0)1865 206206
Blackwell Science, Inc., 350 Main Street,
Malden, MA 02148-5018, USA
    Tel: +1 781 388 8250
Iowa State Press, a Blackwell Publishing
Company, 2121 State Avenue, Ames, Iowa
50014-8300, USA
    Tel: +1 515 292 0140
Blackwell Publishing Asia Pty Ltd,
550 Swanston Street, Carlton South,
Victoria 3053, Australia
    Tel: +61 (0)3 9347 0300
Blackwell Wissenschafts Verlag,
Kurfürstendamm 57, 10707 Berlin, Germany
    Tel: +49 (0)30 32 79 060

This edition published 2003 by Blackwell
Science Ltd

Revised and updated from *Pre-Contract Practice
for the Building Team* (Eighth Edition, © The Aqua
Group, 1960–1992) and *Contract Administration
for the Building Team* (Eighth Edition, © The Aqua
Group 1965–1996)

Library of Congress
Cataloging-in-Publication Data
is available

ISBN 0-632-05485-9

A catalogue record for this title is available from
the British Library

Set in 10/13 pt Palatino
by Sparks Computer Solutions Ltd, Oxford
http://www.sparks.co.uk
Printed and bound in Great Britain by
MPG Books Ltd, Bodmin, Cornwall

For further information on
Blackwell Science, visit our website:
www.blackwell-science.com

Contents

Introduction

Over 20 years ago, the Aqua Group Publications formed part of the reading material for our degree courses. We always found the Aqua Books to be comprehensive, informative and readily understood. It is our hope that, as editors for the new combined title of 'Pre-Contract Practice and Contract Administration for the Building Team', we have maintained the Aqua Group's tradition which dates back to 1960.

Pre-Contract Practice was first published in 1960 and *Contract Administration* was first published in 1965. Both books have been revised over the intervening

Introduction

years, with the latest editions appearing in 1992 and 1996, respectively. Much has changed since, so in this book not only have we brought together all of the material that appeared in those two publications, we have brought it up to date, we have edited out the inevitable duplications and we have added a chapter on capital allowances.

Despite these major revisions, we have aimed to continue the tradition of the Aqua Group and have tried to keep the concepts and terminology simple so the reader quickly grasps the essence of the subject matter. In this context, we have continued to use the terms *architect, quantity surveyors,* etc. without reference to gender and, to save repetition, we have reverted to 'he' universally where the sense allows. This, of course, is not to deny the existence of many able and eminent practitioners complying with the alternative specification! Indeed, we continue to echo the adage *vive la difference.* In a like manner, in the contractual context, we have still not adopted the familiar but long-winded title of *architect/contractor administrator.* We trust that our readers will continue to support the sentiment behind these decisions.

All of the Aqua Group's books have assumed the use of the JCT Standard Form of Building Contract. This book is no different and, except where noted otherwise, assumes the use of the JCT Standard Form of Building Contract

… vive la différence …

1998 Edition incorporating Amendments 1, 2, 3 and 4. It is referred to throughout as JCT98.

Success in building requires that a complicated series of interactions be completed in a logical and pre-determined sequence. The employer will then get the building he wants, at the planned price, in the required time. In this book, we have aimed to set down the principles of good practice to be applied to the pre- and post-contract processes to make this goal more readily achievable.

We discuss those factors that need to be addressed by the building team in developing a project from the earliest involvement of a consultant through to the issue of the final certificate. We also discuss the effect of alternative procurement routes on the appointment and responsibilities of members of the design team. In the last three chapters we consider delays and disputes, capital allowances and insolvency.

We are grateful to the following organisations for permitting us to use their copyright to reproduce some of their standard forms in our examples and extracts from some of their published documents elsewhere in this book:

- **Building Cost Information Service Ltd** for their Standard Form of Cost Analysis elements.
- **Institute of Clerks of Works of Great Britain Incorporated** for their Clerk of Works Project Report form.
- **RIBA Enterprises Ltd** for their RIBA Outline Plan of Work 1998, Uniclass pages 87 to 91 inclusive, Architect's Instruction form, Interim Certificate and Director form, Statement of retention and of Nominated Sub-Contractors' values form, Notification to Nominated Sub-Contractor concerning amount included in certificates form, Certificate of Practical Completion form, Certificate of Making Good Defects form, Final Certificate form and Notification of revision to Completion date form.
- **Royal Institution of Chartered Surveyors** for their Valuation form and Statement of Retention and of Nominated Sub-Contractors' Values form.

Given that we come from a quantity surveying background, it would not have been possible for us to update and edit this publication had we not had expert input from practitioners specialising in other areas of the construction industry. The practitioners concerned and the chapters which they contributed are as follows:

- Chapter 5: Drawings and Schedules – Peter Ullathorne JP, AADipl, RIBA, FRSA, FRSH, AAIA, Vice President, HOK International Limited
- Chapter 6: Specifications – Nick Schumann, Managing Partner of Davis Langdon Schumann Smith
- Chapter 12: Site Duties – Peter Ullathorne JP, AADipl, RIBA, FRSA, FRSH, AAIA, Vice President, HOK International Limited
- Chapter 17: Capital Allowances – Richard Whittaker BSc MRICS, NBW Crosher & James

Special mention, and thanks, must also go to John Quier FRICS, a chartered quantity surveyor in private practice, who was instrumental in carrying out all of the background research which we needed.

In respect of all of the above contributors, we count ourselves fortunate that highly experienced practitioners gave up spare time to let us have the benefit of their advice.

Finally, a debt of gratitude is owed both to the former members of the Aqua Group and to Julia Burden, the Deputy Divisional Director of the Publisher's Professional Division. Our thanks are due to the Aqua Group since they provided us with an excellent foundation on which to take this publication forward and to Julia whose management of the whole process made for a timely publication when, so often, we allowed the distractions of our work commitments to get in the way!

Mark Hackett MA MSc FRICS ACIArb
Ian Robinson BSc(Hons) LLB(Hons) MRICS ACIArb

The Aqua Group

Brian Bagnall
Oliver Burston
John Cavilla
Richard Oakes
Quentin Pickard
Geoffrey Poole
John Townsend
John Willcock
James Williams

Chapter 1
The Building Team

Introduction

Over the years, since the first editions of the Aqua Group's books, the process of building has become more complicated. From inception to completion, through site acquisition, design, contract and construction, each stage has become more time-consuming and thus more expensive. The need to optimise the process is therefore of paramount importance, and the best base from which to achieve this is proper and efficient team working. It is therefore vital that all members of the building team are fully conversant, not only with their own role but also with the roles of others and with the interrelationships at each stage of the project. Each can then play their part fully and effectively, contributing their particular expertise whenever required.

The make-up of any particular building team will depend on the scope and complexity of the project and the contractual arrangement selected. There are already many different methods of managing a project and, no doubt, others will be developed in the future. What follows is set in the context of the JCT98 and, although not exhaustive, will give an indication of the principles involved and the criteria by which other situations can be evaluated.

Parties to a building contract and their supporting teams

The parties to a building contract are the employer and the contractor. Those appointed by these two will complete the *building team* which can include:

The design team

- *employer
- project manager
- *planning supervisor
- *architect

1

- *quantity surveyor
- structural engineer
- building services engineers
- *nominated sub-contractors

The construction team

- *contractor and/or principal contractor
- *site agent (or foreman, described in the contract as the person-in-charge)
- *nominated sub-contractors
- *domestic sub-contractors.

In addition the employer may appoint a:

- *clerk of works.

It should be noted, however, that only those marked with an asterisk are mentioned in JCT98. This list is not exhaustive and to it could be added planners, landscape consultants, process engineers, programmers and the like. Also, some roles may be combined and roles such as the project manager or planning supervisor may be fulfilled by individuals or firms from varying technical backgrounds.

Rights, duties and responsibilities

JCT98 is comprehensive on the subject of the rights, duties and responsibilities of the employer, the contractor and the other members of the building team mentioned in it. Not all the members of the building team are mentioned in JCT98 and those not mentioned will usually be given responsibility by way of delegation from those who are mentioned. This delegation must be spelt out elsewhere in the contract documents – usually in the bills of quantities.

Whatever the size of the building team, each member should be familiar with the contract as a whole and, in particular, with those clauses directly concerning their own work, so that the project can be run smoothly and efficiently. It should be noted that some of the duties are discretionary, some are mandatory and some even have statutory backing.

The important rights, duties and liabilities of individual members of the building team are given in a schedule towards the end of this chapter, by reference to JCT98 Private with Quantities. Other important considerations, relating to members of the building team but not codified in the contract conditions, are set out below.

Named consultants

While the architect and the quantity surveyor are referred to in the contract and are expressly required to perform specific duties (many clauses include the phrase 'the architect shall'), they are not parties to the agreement. Should the contractor have a grievance, if they fail to carry out their duties as defined in the agreement, the only contractual recourse is to seek redress from the employer.

Unnamed consultants with delegated powers

The project manager and the structural or other consulting engineers are not referred to in the contract, nor do they have any powers under the contract. Their position within the building team depends on the agreement they have with the employer or the architect. Where they have been given responsibility by way of delegation, perhaps for design or site inspection, they should be named in the contract documents and the extent of their delegated responsibility should be identified so that they have contractual recognition. Since they have no powers under the contract, if they need to issue instructions this must be done through the architect.

As with the named consultants, if the contractor has any grievance against unnamed consultants the only contractual recourse is to seek redress from the employer.

The project manager

The project manager, who may be considered as the employer's representative, is likely to be appointed at the outset by the employer to whom he is directly accountable. The project manager is responsible for the programming, monitoring and management of the project in its broadest sense, from inception to completion, to ensure a satisfactory outcome. This will involve giving advice to the employer on all matters relating to the project, including the appointment of the architect, quantity surveyor and other consultants. However, not being mentioned in the contract, the project manager's position in respect of the contract works, which will be only a part of his overall duties and responsibilities, must be clearly determined and described in the contract documents. The duties and responsibilities delegated must not exceed those set out in the contract in respect of the employer.

The planning supervisor

The Construction (Design and Management) Regulations 1994 (CDM94), which came into effect on 31 March 1995, introduced the *planning supervisor* to the building team. The planning supervisor is specifically referred to in JCT98 as being appointed by the employer, pursuant to regulation 6(5) of the CDM Regulations and, under Article 6.1, the planning supervisor is stated to be the architect, unless some other individual is identified.

The planning supervisor is required to:

- ensure that the design avoids foreseeable risks
- ensure co-operation between designers
- be able to advise the client (as the employer is termed in CDM94) and contractor on the competence of appointed designers
- be able to advise the client on the competence of a contractor
- ensure preparation of a draft health and safety plan for the contractors at tender stage
- give notice of the project to the Health and Safety Executive
- ensure preparation of the health and safety file
- review and amend the health and safety file as work proceeds
- deliver the health and safety file to the employer on completion of the project.

The principal contractor

In JCT98, under Article 6.2, the principal contractor is stated to be the contractor or such other contractor as the employer shall appoint as the principal contractor pursuant to regulation 6(5) of CDM94. The employer must be reasonably satisfied that such principal contractor is competent and can allocate sufficient resources to ensure compliance with the regulations. The definition of a principal contractor in CDM94 allows for considerable flexibility, but for the majority of construction contracts carried out under JCT98 the main contractor is appointed as the principal contractor.

Since there must always be a principal contractor in post while work is in progress on site, problems could arise where enabling works contracts, such as for demolition or piling, are required before the commencement of the main contract, or where a fitting-out contract follows completion of the main contract. In these circumstances the employer will probably appoint a succession of principal contractors but care will need to be taken to ensure responsibility passes properly from one to the next.

The principal contractor is required to:

- ensure that the other contractors comply with their obligations on health and safety matters
- co-ordinate co-operation between sub-contractors
- ensure compliance with the health and safety plan
- ensure that all relevant persons comply with the health and safety plan
- ensure that only authorised persons are allowed on site
- display a copy of the notice submitted to the Health and Safety Executive
- provide the planning supervisor with the information he requires for the health and safety file
- instruct other contractors so that the principal contractor complies with his duties.

Nominated sub-contractors

Nominated sub-contractors feature as members of the design team and as members of the construction team. This is because, as well as carrying out work on site, they are often involved in the design and planning of specialist works in advance of the appointment of the main contractor. Because of this special relationship, nominated sub-contractors acquire rights under JCT98 which are not afforded to domestic sub-contractors, but by the same token they have responsibilities which must be discharged in close liaison with the rest of the building team. In JCT98, nominated sub-contractors' special rights to payment are safeguarded by imposing duties on the quantity surveyor, the architect and the contractor.

The clerk of works

The clerk of works is normally appointed by the employer to act, under the direction of the architect, solely as an inspector of the works. Traditionally, he would have been an experienced tradesman, perhaps a carpenter and joiner or bricklayer. However, with today's highly complex and high-tech buildings, the architect, who will normally recommend the appointment, may need a technically experienced or qualified person and here the Institute of Clerks of Works will be able to assist in finding the right person. The clerk of works should be ready to take up his duties before the date of possession (how early will depend on the size and complexity of the project) and will either be resident on site or will visit the site on a regular basis during the period of the works.

Statutory requirements

Under JCT98, the parties and the various consultants are charged with many duties, some of which are discretionary (the architect may …) and some of which are mandatory (the contractor shall …). In addition, provision is made for compliance with duties and responsibilities imposed by legislation (statutory matters) as follows:

- clause 6   Statutory obligations, notices, fees and charges
- clause 6A   CDM Regulations
- clause 15   Value Added Tax
- clause 31   Construction Industry Scheme.

These may change from time to time and may be augmented by the addition of regulations issued by government under *enabling Acts*. Constant vigilance is therefore required to keep up to date.

The CDM Regulations

The CDM Regulations place duties on clients, planning supervisors, designers and principal contractors to plan, co-ordinate and manage health and safety throughout all stages of a construction project. New laws on health and safety were not introduced by the regulations, which merely codify the administrative machinery to provide an audit trail to demonstrate accountability for the management of health and safety throughout the project. It is worthy of note that contravention of the regulations can be a criminal offence. The regulations apply to most projects, but there are a few exceptions – notably, they do not apply to:

- construction work when the local authority is the enforcing authority for health and safety purposes
- work which is expected to last no more than 30 days or involve no more than 500 person days of construction work and involves less than five people on site at any one time – apart from Regulation 13 (designer duties)
- work for a domestic client, unless a developer is involved – apart from Regulation 7 (site notification) and Regulation 13 (designer duties).

However, where the construction phase of a project involves demolition, CDM94 applies regardless of whether or not the work would otherwise have been exempt.

Under the overall guidance of the planning supervisor, it is the designer who is the responsible person in the early stages of a project, and, while such duties will obtain throughout the design and construction periods, it is in the pre-contract stages that the majority of the designer's responsibilities under CDM94 are to be discharged. It is the designer's duty to tell the client initially what the client's duties are. The designer should design to reduce, if not avoid, as far as reasonably practicable, risks to health and safety, both during construction and during subsequent occupation, and ensure that the design documentation includes adequate information on health and safety. Such information should be included on drawings or in specifications and should be passed to the planning supervisor for inclusion in the health and safety file. The designer should not get involved with the principal contractor's duties – CDM94 is quite clear that the management of safety during construction is the responsibility of the principal contractor.

Avoiding disputes

In *Tenders and Contracts for Building*, reference is made to the importance of formal procedures and proper documentation in the efficient and smooth running of building contracts – clarity and transparency are perhaps the key words. This is particularly relevant in the avoidance of disputes. As projects become more complex, costs increase, margins tighten and employers demand greater quality and financial control, the margin between success and failure narrows, and there is less flexibility to absorb the *swings and roundabouts* which have ever been a feature of the construction industry. The likelihood of disputes arising is, therefore, increased.

The building team must be vigilant and ensure that the procedures employed, and the working relationships built up, produce an environment of co-operation rather than discord. Just as the building team strives for perfection in its working relationships, so the JCT strives to refine its contracts to take account of current practice and the latest decisions in the courts. Hence the frequent revisions to its standard forms of building contract, which aim to achieve clarity and certainty in formalising the relationships between the parties.

In the event that disputes do arise, the importance of proper documentation and compliance with formal procedures cannot be over-stressed. If dispute proceedings become inevitable, it should be some comfort to know that proper documentation will be an asset rather than a liability.

On the assumption that the contract documents are complete, that tenders are reasonable and reflect current prices, and that information is available when required and not subject to late changes on the part of the employer or the architect, the origin of contractual disputes is seldom found to be in the dishonesty or incompetence of any party but rather in the failure of one member of the team to convey information clearly to another. Unfortunately, what is clear in the mind of the architect, for example, might be *misty* to the quantity surveyor and *foggy* to the contractor. This can lead to all sorts of problems! Conveying intentions and instructions clearly is vital in the successful management of a contract and, all too often, it is the breakdown or failure of communications that is the root cause of a dispute.

Communications

Many books, conferences and papers have been devoted to the subject of *communications* and it is a matter that cannot be dealt with exhaustively here. However, set out below are certain golden rules to be observed by all members of the team in their dealings with each other:

- Do not (wherever possible) tamper with the standard clauses of JCT98 – if, however, the client insists upon it, employ a specialist.
- Ensure that the contract is executed prior to any start on site.
- Ensure that all team members have certified copies of the contract documents.
- Be realistic when completing the appendix to the contract and identify such information in the bills of quantities at tender stage.
- Issue all instructions to the contractor through the architect.
- Issue all instructions to sub-contractors or suppliers through the contractor.
- Use standard forms or formats for all routine matters such as instructions, site reports, minutes of meetings, valuations and certificates, preferably with sequential numbering.
- If instructions are issued other than in writing, ensure that they are confirmed in writing as soon as possible after the event.
- Ensure that everybody is kept informed, not just those who have to act.
- Be precise and unambiguous.
- Act promptly.

Examples of standard forms and suggested standard layouts for the more important communications passing between members of the building team are given in later chapters.

Schedule of important rights, duties and liabilities of members of the building team based on JCT98 Private with Quantities

The employer

Recitals

First	Duty to state the nature of intended works.
Third	Duty to sign the contract drawings and the contract bills.
Fourth	Duty to give status for the purposes of the Construction Industry Scheme.
Sixth	Optional duty to provide an information release schedule.
Seventh	Duty to give terms applicable to any bonds required.

Articles

2	Duty to pay the contract sum.
3	Duty to nominate and appoint a replacement architect.
4	Duty to nominate and appoint a replacement quantity surveyor.
6.1	Duty to appoint a replacement planning supervisor.
6.2	Duty to appoint a replacement principal contractor.

Attestation

	Duty to sign.

Clauses

1.6	Duty to notify the contractor of the appointment of a replacement planning supervisor or a replacement principal contractor.
1.9	Duty to notify the contractor of the appointment of an employer's representative.
4.1	Right to employ others if the contractor does not comply with the architect's instructions.
5.1	Duty to retain custody of the contract documents (Local Authority edition only).

5.4.1	Right to agree a variation to the information release schedule.
5.7	Duties in relation to the confidential nature of the contract documents.
5.8	Right to be issued with all architect's certificates.
5.9	Right to receive, from the contractor, information on and 'as built' drawings of performance specified work including its maintenance and operation.
6.2	Right to be indemnified, by the contractor, against liability for statutory fees or charges.
6A.1	Duties in relation to the planning supervisor and the principal contractor under the CDM Regulations.
6A.2	Duty to notify the planning supervisor and the architect of any amendments, by the contractor, to the health and safety plan.
7	Right to consent to the architect instructing that setting out errors shall not be corrected.
8.4	Right to agree to allow non-compliant materials or goods to remain.
9.1	Right to indemnification in respect of claims relating to the infringement of patent rights.
12	Right to appoint a clerk of works.
13.1	Right to impose or vary any obligations or restrictions in regard to access to or use of the site, working space, working hours and the order of work.
13.4.1.1	Right to agree the method to be adopted for the valuation of variations.
13.4.1.2.A4.3	Right to refer the contractor's price statement and the amended price statement as a dispute or difference to the adjudicator.
13.4.1.3	Right to approve an alternative method of valuing variations to nominated sub-contract works.
13A.3.1	Rights and duties in respect of the acceptance of a 13A Quotation.
13A.4	Right to not accept a 13A Quotation.
13A.6	Duty not to use a 13A Quotation that has not been accepted.
13A.7	Right to agree to vary the stated time periods associated with a 13A Quotation.
16.1	Right to ownership of unfixed materials and goods on-site.
16.2	Right to ownership of unfixed materials and goods off-site.
17.2 & 17.3	Right to consent to the architect instructing that defects, shrinkages and other faults shall not be made good.
18.1	Right to take partial possession of the works with the consent of the contractor.
18.1.3	Duty to insure the relevant part on taking partial possession of the works.
19.1.1	Right to assign the contract with the consent of the contractor.
19.1.2	Right, after practical completion, to assign the right to bring proceedings in the name of the employer.

19.3.2	Right to add to the list of named persons with the consent of the contractor.
19.4.2.2	Right to ownership of domestic sub-contract unfixed materials and goods on site.
20 & 21	Rights, duties and liabilities in respect of injury to or the death of persons and loss of or damage or injury to property real or personal and related insurance.
22.3	Duties in respect of the joint names policy of insurance.
22A	Rights and duties when insurance of the works is the responsibility of the contractor.
22B	Duty to insure the works as an alternative to clause 22A.
22C	Duty to insure the works, the existing structures and their contents when the works are in or are extensions to existing buildings.
22D	Right to require the contractor to insure for loss of liquidated and ascertained damages due to extensions of time for delays caused by the specified perils.
22FC	Rights, duties and liability in respect of the Joint Fire Code.
23.1.1	Duty to give the contractor possession of the site.
23.1.2	Right to defer giving the contractor possession of the site.
23.3.2 & 23.3.3	Right to use or occupy the site or the works with the consent of the contractor and the duty to pay for any related increase in insurance premiums.
24.2	Rights and duties in respect of liquidated and ascertained damages.
25.4.12, 26.4.2 & 28.2.3	Liabilities in the event of delays caused by the execution of work or by the supply of materials or goods not forming part of the contract.
25.4.12, 26.2.6 & 28.2.4	Liabilities in the event of a failure to give timely ingress to or egress from the site.
25.4.13	Liability in the event of a deferment of giving possession of the site.
25.4.17 & 26.2.9	Liabilities in the event of compliance or non-compliance with the CDM Regulations.
25.4.19 & 26.2.11	Liability in respect of any impediment, prevention or default whether by act or omission.
27.2	Rights and duties in respect of determination of the employment of the contractor in the event of a default by the contractor.
27.3	Rights and duties in respect of determination of the employment of the contractor in the event of the contractor's bankruptcy or insolvency.
27.4	Right to determine the employment of the contractor in the event of corruption by the contractor.
27.5	Optional rights and duties in the event of the contractor's insolvency.
27.6	Rights and duties in respect of the employment of other persons to carry out and complete the works following determination of the employment of the contractor.

27.7	Rights and duties if the employer decides not to proceed with the works following determination of the employment of the contractor.
27.8	Right to retain common law rights and remedies in the event of determination by the employer.
28.2.3 & 28.2.4	Liability in the event of continuing or repeating a specified default or suspension event.
28.3	Duties in the event of bankruptcy or insolvency of the employer.
28.4	Rights, duties and liabilities in the event of determination of the employment of the contractor as a result of the employer's default, bankruptcy or insolvency.
28A	Rights, duties and liabilities in the event of determination of the employment of the contractor when neither party is at fault.
29	Rights and duties in respect of work on site not forming part of the contract.
30A	Duty under the Construction Industry Scheme if a 'contractor'.
30.1	Rights, duties and liabilities in respect of payments due to the contractor.
30.3	Rights and duties in respect of payment for materials or goods off-site.
30.4	Right to deduct and retain the retention from interim payments.
30.4A	Rights in respect of the provision of a bond, by the contractor, in lieu of retention.
30.5	Rights and duties in respect of the retention.
30.8	Rights, duties and liabilities in respect of the final payment.
31	Rights, duties and liabilities in respect of the Construction Industry Scheme.
34	Rights and liabilities in the event of fossils, antiquities and other objects of interest or value being found on site.
35.1.4	Right to agree the nomination of a sub-contractor.
35.6.1	Duty to sign NSC/T part 2.
35.6.2	Right to enter into Agreement NSC/W.
35.13.5	Duty to pay nominated sub-contractors direct in the event of the contractor's failure to pay.
35.13.6	Rights and duties in respect of pre-nomination payments to nominated sub-contractors.
35.18	Rights and duties if a nominated sub-contractor fails to remedy defects after early final payment.
35.20	Liability to nominated sub-contractors.
35.22	Rights where the liability of the nominated sub-contractor to the contractor is limited.
35.24	Rights and duties in circumstances where re-nomination is necessary.
36.5.1	Rights when the contract of sale between the contractor and a nominated supplier restricts, limits or excludes the liability of the nominated supplier to the contractor.
41A	Rights and duties in respect of adjudication.

41B   Rights and duties in respect of arbitration.
42.12   Right to agree with the contractor that the architect may instruct the provision of performance specified work as a variation.

Supplemental provisions

Rights, duties and liabilities in respect of VAT.

The contractor

Recitals

Second   Duty to provide a fully priced copy of the bills of quantities and the optional duty to provide a priced activity schedule.
Third   Duty to sign the contract drawings and the contract bills.

Articles

1   Duty to carry out and complete the works.
3   Right to object to the appointment of any person nominated as a replacement architect.
3   Right to object to the appointment of any person nominated as a replacement quantity surveyor.

Attestation

Duty to sign.

Clauses

1.5   Liability to carry out and complete the works in accordance with the conditions.
1.6   Right to be notified of the appointment of a replacement planning supervisor or a replacement principal contractor.
1.9   Right to be notified of the appointment of an employer's representative.
2.1   Duty to carry out and complete the works in compliance with the contract documents.
2.3 & 2.4 Duties in the event of finding any discrepancies in or divergences between documents.
4.1   Rights, duties and liabilities in respect of architect's instructions.
4.2   Right to request the architect to state which provision of the conditions empowers him to issue any particular instruction.

4.3.2	Rights and duties in connection with architect's oral instructions.
5.1	Right to inspect the contract documents.
5.2	Right to receive, from the architect, contract documents, drawings and bills of quantities.
5.3.1.1	Right to receive, from the architect, descriptive schedules and the like.
5.3.1.2	Duty to provide and update the master programme.
5.4.1	Right to agree a variation to the information release schedule.
5.4.2	Rights and duties in respect of the provision of such further drawings and details and the issue of such instructions as are necessary to enable the contractor to carry out and complete the works.
5.5	Duty to keep documents on site.
5.6	Duty to return documents, bearing the name of the architect, to the architect, if requested.
5.7	Duties in relation to the confidential nature of the contract documents.
5.8	Right to be issued with a copy of all architect's certificates.
5.9	Duty to supply information on and 'as built' drawings of performance specified work including its maintenance and operation.
6.1	Rights and duties in respect of statutory requirements.
6.2	Duty to indemnify the employer against liability for statutory fees or charges.
6.3	Duty to comply with the reasonable requirements of the principal contractor.
6A.2	Duty to comply with all the duties of a principal contractor, set out in the CDM Regulations, while the contractor is the principal contractor.
6A.4	Duty to provide the information necessary for the preparation of the health and safety file.
7	Rights and duties in respect of levels and setting out of the works.
8.1	Duty to comply with the required standards of materials, goods and workmanship.
8.2	Duty to provide proof of compliance of materials and goods and the right to be informed of unsatisfactory work within a reasonable time.
8.3	Rights, duties and liabilities in respect of work opened up for inspection and the testing of any materials or goods.
8.4	Rights, duties and liabilities in respect of work not in accordance with the contract.
8.5	Right to be consulted in the event of a failure to carry out work in accordance with the health and safety plan.
9.1	Rights and liability in respect of claims relating to the infringement of patent rights.
10	Duty to keep a competent person-in-charge upon the site.
11	Duty to provide access for the architect to the works and workshops or other places where work is being prepared for the contract.
12	Duty to afford the clerk of works every reasonable facility.
13.2.3	Right to disagree with the application of clause 13A.

13.4.1.1	Right to agree the method to be adopted for the valuation of variations.
13.4.1.2.A1	Right to submit a price statement to the quantity surveyor.
13.4.1.2.A2	Right to be notified, by the quantity surveyor, that a price statement is, or is not, accepted (in whole or in part).
13.4.1.2.A4	Rights and duties when a price statement is not accepted.
13.4.1.2.A5	Right to refer a contractor's price statement as a dispute or difference to the adjudicator.
13.4.1.2.A7	Right to be notified as to whether or not any amount required to be paid in lieu of ascertainment of direct loss and/or expense or any adjustment required to the time for completion of the works, attached to the contractor's price statement, is accepted.
13.4.1.3	Right to agree an alternative method of valuing variations to nominated sub-contract works.
13.5	Duties in respect of daywork vouchers.
13.6	Right to be given the opportunity of being present and of taking notes, at the time of taking measurements for the valuation of variations.
13A.1	Duties in respect of 13A Quotations.
13A.3.1	Right to receive, from the employer, notification of the acceptance of a 13A Quotation.
13A.3.2	Right to receive, from the architect, confirmation of acceptance of a 13A Quotation.
13A.4	Rights when the employer does not accept a 13A Quotation.
13A.6	Duty not to use a 13A Quotation that has not been accepted.
13A.7	Right to agree to vary the stated time periods associated with a 13A Quotation.
15.3	Rights when the employer becomes exempt from VAT.
16.1	Liability for unfixed materials and goods on site.
16.2	Liability for unfixed materials and goods off-site.
17.2 to 17.5	Rights, duties and liabilities in respect of defects, shrinkages and other faults.
18.1	Rights, duties and liabilities in the event of the employer taking partial possession of the works.
19.1.1	Right to assign the contract with the consent of the employer.
19.2 to 19.4	Rights, duties and liabilities in respect of sub-letting.
19.5	Rights and liabilities in respect of the works of nominated sub-contractors.
20 & 21	Rights, duties and liabilities in respect of injury to or the death of persons and loss of or damage or injury to property real or personal and related insurance.
22.3	Duties in respect of the joint names policy of insurance.
22A	Rights and duties in respect of insurance of the works.
22B	Rights and duties when insurance of the works is the responsibility of the employer.

22C	Rights and duties when the employer is required to insure the works, the existing structures and their contents.
22D	Rights and duties in the event that insurance is required for loss of liquidated and ascertained damages due to extensions of time for delays caused by the specified perils.
22FC	Rights and duties in respect of the Joint Fire Code.
23.1	Right to be given possession of the site and a duty to begin and to regularly and diligently proceed with the works.
23.3.2 & 23.3.3	Rights and duties if the employer seeks consent to use or occupy the site or the works.
24.2	Rights and liabilities in respect of liquidated and ascertained damages.
25.2	Duties when the progress of the works is being or is likely to be delayed.
25.3.1 & 25.3.3	Right to be given an extension of time.
25.3.4	Duty to constantly use best endeavours to prevent delay in the progress of the works.
25.4	Rights and duties in respect of relevant events.
26.1	Rights and duties when the contractor has incurred or is likely to incur direct loss and/or expense due to deferment of being given possession of the site or because the regular progress of the works has been materially affected by a listed matter.
26.3	Right to be notified of extensions of time given in respect of some relevant events.
26.4	Rights and duties in respect of claims from nominated sub-contractors for reimbursement of loss and/or expense.
26.6	Right to retain common law rights and remedies in respect of loss and/or expense.
27.2	Rights in respect of determination of the employment of the contractor in the event of a default by the contractor.
27.3	Duty in the event of the bankruptcy or insolvency of the contractor.
27.5	Rights and liabilities in the event of the contractor's insolvency.
27.6	Rights and liabilities in the event of the employment of other persons to carry out and complete the works following determination of the employment of the contractor.
27.7	Rights and liabilities if the employer decides not to proceed with the works following determination of the employment of the contractor.
28.2	Rights and duties in respect of determination of the employment of the contractor in the event of a default by the employer.
28.3.3	Right to determine the employment of the contractor in the event of the employer's bankruptcy or insolvency.
28.4	Rights and duties in the event of determination of the employment of the contractor as a result of the employer's default, bankruptcy or insolvency.

28.5	Right to retain common law rights and remedies in the event of determination by the contractor.
28A	Rights, duties and liabilities in the event of determination of the employment of the contractor when neither party is at fault.
29	Rights and duties in respect of work on site not forming part of the contract.
30.1	Rights, duties and liabilities in respect of payments due from the employer.
30.3	Rights, duties and liabilities in respect of payment for materials or goods off-site.
30.4A	Duties in respect of the provision of a bond, by the contractor, in lieu of retention.
30.5	Rights in respect of the retention.
30.6	Rights and duties in respect of the final adjustment of the contract sum.
30.8	Rights, duties and liabilities in respect of the final certificate.
31	Rights, duties and liabilities in respect of the Construction Industry Scheme.
34	Rights and duties in the event of fossils, antiquities and other objects of interest or value being found on site.
35.1.4	Right to agree the nomination of a sub-contractor.
35.2	Right to tender for nominated sub-contractor work.
35.5	Right to object to a proposed nominated sub-contractor.
35.6	Right to receive, from the architect, an instruction on Nomination NSC/N nominating a sub-contractor.
35.7 to 35.9	Rights and duties upon receipt of an instruction nominating a sub-contractor.
35.13	Rights, duties and liabilities in respect of payments due to nominated sub-contractors.
35.14	Duties in relation to extension of the period or periods for completion of nominated sub-contract works.
35.15	Rights and duties in the event of a failure to complete nominated sub-contract works.
35.17 to 35.19	Rights, duties and liabilities in respect of early final payment of nominated sub-contractors.
35.21 & 35.22	Liability in respect of nominated sub-contract works.
35.24	Rights and duties in circumstances where re-nomination is necessary.
35.25	Rights and duties in respect of the determination of the employment of a nominated sub-contractor.
36.3.2	Right to have expense properly incurred in obtaining materials or goods from a nominated supplier added to the contract sum.
36.4	Rights, duties and liabilities to be provided for in the terms of the contract of sale applicable to a nominated supplier.

38	Rights, duties and liabilities in respect of contribution, levy and tax fluctuations.
39	Rights, duties and liabilities in respect of labour and materials cost and tax fluctuations.
40	Rights, duties and liabilities in respect of fluctuations calculated by use of price adjustment formulae.
41A	Rights and duties in respect of adjudication.
41B	Rights and duties in respect of arbitration.
42	Rights, duties and liabilities in respect of performance specified work.

Supplemental provisions

Rights, duties and liabilities in respect of VAT.

The architect

Articles

3	Duty not to disregard or overrule any certificate or opinion or decision or approval or instruction given or expressed by a preceding architect.

Clauses

2.3 & 2.4	Duty to issue instructions in regard to discrepancies in or divergences between documents.
4.1.2	Right to issue a notice requiring compliance with an instruction.
4.2	Duty to respond to a request from the contractor to state which provision of the conditions empowers the issue of any particular instruction.
4.3.1	Duty to issue all instructions in writing.
4.3.2	Duties in connection with oral instructions.
5.1	Duty to retain custody of the contract documents (private edition only).
5.2	Duty to provide the contractor with contract documents, drawings and bills of quantities.
5.3.1.1	Duty to provide the contractor with descriptive schedules and the like.
5.3.1.2	Right to receive, from the contractor, the contractor's master programme and updates of it.
5.4.1	Duty to provide information in accordance with the information release schedule.
5.4.2	Rights and duties in respect of the provision of such further drawings and details and the issue of such instructions as are necessary to enable the contractor to carry out and complete the works.

5.5	Right to have documents available on site.
5.6	Right to request the contractor to return documents which bear the name of the architect.
5.7	Duties in relation to the confidential nature of the contract documents.
5.8	Duty to issue certificates to the employer with copies to the contractor.
6.1	Rights and duties in respect of statutory requirements.
6A.2	Right to be notified, by the employer, of any amendments, by the contractor, to the health and safety plan.
7	Rights and duties in respect of levels and setting out of the works.
8.1.4	Right to consent to the substitution of materials or goods for use in performance specified work.
8.2	Right to require proof of the standard of materials and goods and the duty to express any dissatisfaction in respect of materials, goods or workmanship within a reasonable time from the execution of the unsatisfactory work.
8.3	Right to issue instructions requiring the opening up of work for inspection and the carrying out of tests on any materials and goods.
8.4	Rights and duties in respect of work not in accordance with the contract.
8.5	Rights and duties in the event of a failure to carry out work in a proper and workmanlike manner and/or in accordance with the health and safety plan.
8.6	Right to exclude persons from the site.
11	Right to have access to the works and workshops or other places where work is being prepared for the contract.
12	Right to direct the clerk of works and to confirm any direction of the clerk of works properly given to the contractor.
13.2.1	Right to issue instructions requiring a variation.
13.2.3	Right to instruct a variation when the contractor disagrees with the application of clause 13A.
13.2.4	Right to sanction any variation made by the contractor.
13.3	Duty to issue instructions in regard to the expenditure of provisional sums.
13.4.1.2.A2, 13.4.1.2.A4.1 & 13.4.1.2.A7.1	Rights to be consulted by the quantity surveyor about the contractor's price statement.
13.5	Right to receive, from the contractor, daywork vouchers for verification.
13A.3.2	Duty to confirm acceptance of a 13A Quotation.
13A.4	Duty, in the event that a 13A Quotation is not accepted, to instruct that the variation is, or is not, to be carried out.
16.1	Right to consent to the removal from site of unfixed materials and goods.
17.1	Duty to issue a certificate of practical completion.

17.2 to 17.5	Rights and duties in respect of defects, shrinkages and other faults.
18.1	Rights and duties in the event of the employer taking partial possession of the works.
19.2.2	Right to consent to the contractor sub-letting part of the works.
19.3.2	Right to act on behalf of the employer in respect of adding to the list of named persons.
21.1.2	Right to receive, from the contractor, documentary evidence that insurances have been taken out and are being maintained.
21.2	Right to instruct the contractor to take out a policy of insurance in respect of liability, etc. of the employer and to receive the policies and premium receipts therefore.
22A	Rights and duties when insurance of the works is the responsibility of the contractor.
22B	Rights and duties when insurance of the works is the responsibility of the employer.
22C	Rights and duties when the employer is required to insure the works, the existing structures and their contents.
22D	Rights and duties in the event that insurance is required for loss of liquidated and ascertained damages due to extensions of time for delays caused by the specified perils.
22FC	Rights and duties in respect of the Joint Fire Code.
23.2	Right to issue instructions in regard to the postponement of work.
24.1	Duty to issue certificates in the event of the contractor failing to complete the works by the completion date.
25.2	Right to receive, from the contractor, notices that the progress of the works is being or is likely to be delayed.
25.3	Rights and duties in respect of fixing a new completion date.
26.1	Rights and duties when the contractor has incurred or is likely to incur direct loss and/or expense due to deferment of being given possession of the site or because the regular progress of the works has been materially affected by a listed matter.
26.3	Duty to notify the contractor of extensions of time given in respect of some relevant events.
26.4	Rights and duties in respect of claims from nominated sub-contractors for reimbursement of loss and/or expense.
27.2.1	Right to notify the contractor of defaults.
27.6	Rights and duties in the event of the employment of other persons to carry out and complete the works following determination of the employment of the contractor.
28.2.2.4	Rights in respect of a failure by the employer to give timely ingress to or egress from the site.
30.1	Rights and duties in respect of interim certificates.
30.3	Rights in respect of payments for materials off-site.

30.4A.1	Duty to prepare or instruct the quantity surveyor to prepare a statement of retention.
30.5	Rights and duties in respect of the retention.
30.6	Rights and duties in respect of the final adjustment of the contract sum.
30.7	Duty to issue an interim certificate to include finally adjusted nominated sub-contract sums.
30.8	Duty to issue the final certificate and to inform all nominated sub-contractors of the date of its issue.
34	Rights and duties in the event of fossils, antiquities and other objects of interest or value being found on site.
35.1.4	Right to agree the nomination of a sub-contractor on behalf of the employer.
35.2.1	Right to consent to sub-letting.
35.5.2	Rights and duties when the contractor objects to a proposed nominated sub-contractor.
35.6	Duties in respect of nominating a sub-contractor.
35.7.2	Right to receive, from the contractor, the completed Agreement NSC/A and the agreed and signed NCS/T Part 3.
35.8 & 35.9	Rights and duties in the event of non-compliance with a nomination instruction.
35.13	Rights and duties in respect of payments to nominated sub-contractors.
35.14	Duties in relation to extension of the period or periods for completion of nominated sub-contract works.
35.15	Rights and duties in the event of a failure to complete nominated sub-contract works.
35.16	Duty to issue a certificate of practical completion of the works of a nominated sub-contractor and to send a copy to the nominated sub-contractor.
35.17 & 35.18	Rights and duties in respect of early payment of nominated sub-contractors.
35.24	Rights and duties in circumstances where a re-nomination is necessary.
35.25	Duties in relation to determination of the employment of a nominated sub-contractor.
36.2	Duty to issue instructions for the purpose of nominating a supplier.
36.3.2	Right to add expense, properly incurred by the contractor in obtaining materials or goods from a nominated supplier, to the contract sum.
36.4 & 36.5	Rights and duties in respect of the nomination of a supplier.
38	Rights in respect of contribution, levy and tax fluctuations.
39	Rights in respect of labour and materials cost and tax fluctuations.
40	Duties in respect of fluctuations calculated by use of price adjustment formulae.

42   Rights and duties in respect of performance specified work.

Supplemental provisions

            Rights and duties in respect of VAT.

The quantity surveyor

Clauses

5.1   Duty to retain custody of the contract documents (private edition only).

5.7   Duties in relation to the confidential nature of the contract documents.

13.4.1.2.A1   Right to receive, from the contractor, the contractor's price statement.

13.4.1.2.A2   Duty to notify the contractor that a price statement is, or is not, accepted (in whole or in part), after consultation with the architect.

13.4.1.2.A4   Duty when a price statement or part thereof has not been accepted.

13.4.1.2.A7   Duty when the contractor has attached to a price statement the amount he requires to be paid in lieu of ascertainment of direct loss and/or expense or any adjustment he requires to the time for the completion of the works.

13.4.1.2   Alternative B Duty to value variations.

13.6   Duty to give the contractor the opportunity of being present and of taking notes, at the time of taking measurements for the valuation of variations.

13A.1.2   Right to receive, from the contractor, 13A Quotations.

13A.8   Duty in respect of valuing variations to work for which a 13A Quotation has been accepted.

26.1   Duty, if so instructed by the architect, to ascertain the direct loss and/or expense incurred or likely to be incurred by the contractor due to deferment of being given possession of the site or because the regular progress of the works has been materially affected by a listed matter.

26.1.3   Right to receive, from the contractor, details of loss and/or expense (if requested).

26.4.1   Duty, if so instructed by the architect, to ascertain loss and/or expense incurred by a nominated sub-contractor.

30.1.2.1   Duty to make interim valuations whenever the architect considers them to be necessary for the purpose of interim certificates. (When clause 40 applies this duty is not at the discretion of the architect – a valuation is then obligatory prior to each interim certificate).

30.1.2.2	Rights and duties in the event of the contractor submitting an application prior to an interim certificate.
30.4A.1 & 30.5.2.1	Duty, if so instructed by the architect, to prepare a statement of retention.
30.6	Rights and duties in respect of the final adjustment of the contract sum.
34.3.2	Duty, if so instructed by the architect, to ascertain the amount of loss and/or expense incurred by the contractor in the event of fossils, antiquities and other objects of interest or value being found on site.
38	Rights in respect of contribution, levy and tax fluctuations.
39	Rights in respect of labour and materials cost and tax fluctuations.
40	Rights in respect of fluctuations calculated by use of price adjustment formulae.

Planning supervisor

Clause

6A.2	Right to be notified, by the employer, of any amendments, by the contractor, to the health and safety plan.
6A.4	Rights in respect of the information necessary for the preparation of the health and safety file.
42.2	Rights in respect of performance specified work.

Clerk of works

Clause

11	Right to give directions to the person-in-charge.
12	Duty to act solely as inspector on behalf of the employer under the direction of the architect.
34.1	Right to be informed, by the contractor, of the discovery on site and the precise location of fossils, antiquities and other objects of interest or value.

Person-in-charge

Clause

| 10 | Duty to receive instructions of the architect and directions of the clerk of works. |

Sectional completion and contractor's designed portion supplements

The sectional completion supplement is for incorporation into JCT98 when completion of separate sections of the works is required at different times. The contractor's designed portion supplement is for incorporation into JCT98 when the contractor is required to design part of the works. When either or both of these supplements are used, the additional duties, rights and liabilities of the building team are clearly defined.

Chapter 2
Assessing the Needs

The structure

The successful development of any building project requires the acceptance by all parties of an underlying structure or framework of operation. For a traditional building project the Outline Plan of Work published by RIBA Publications and reproduced as Fig. 2.1 is an excellent model framework for running a project. It also provides a useful checklist for key activities.

As work proceeds, it is prudent to reconsider earlier decisions to ensure that the developing design continues to satisfy the employer's overall criteria. Generally, the procedures recommended in this book follow the Outline Plan of Work. However, the plan can be modified to suit those procedures required by alternative methods of building procurement. It is worthy of note that RIBA Publications also publishes a separate Outline Plan of Work for use on design and build projects.

The brief

The initial brief from the employer will often be little more than a statement of intent; it may be no more than a telephone call. At this stage there is unlikely to be any formal appointment. After the initial euphoria surrounding *getting a job,* members of the design team will quickly realise that far more information is required from both the potential employer and from within their own organisations.

From the potential employer each member of the design team will, typically, require details of:

- whether or not they are in competition with others for the commission
- the other members of the design team
- the status of the employer – contractor, developer, purchaser or owner occupier

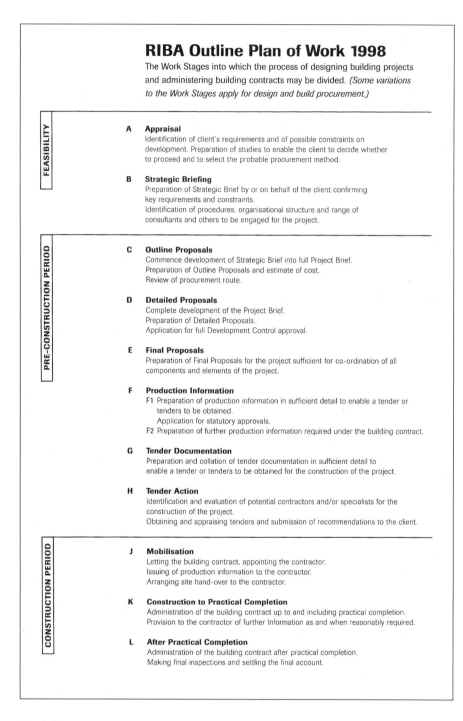

RIBA Outline Plan of Work 1998

The Work Stages into which the process of designing building projects and administering building contracts may be divided. *(Some variations to the Work Stages apply for design and build procurement.)*

FEASIBILITY

A   Appraisal
Identification of client's requirements and of possible constraints on development. Preparation of studies to enable the client to decide whether to proceed and to select the probable procurement method.

B   Strategic Briefing
Preparation of Strategic Brief by or on behalf of the client confirming key requirements and constraints.
Identification of procedures, organisational structure and range of consultants and others to be engaged for the project.

PRE-CONSTRUCTION PERIOD

C   Outline Proposals
Commence development of Strategic Brief into full Project Brief.
Preparation of Outline Proposals and estimate of cost.
Review of procurement route.

D   Detailed Proposals
Complete development of the Project Brief.
Preparation of Detailed Proposals.
Application for full Development Control approval.

E   Final Proposals
Preparation of Final Proposals for the project sufficient for co-ordination of all components and elements of the project.

F   Production Information
F1 Preparation of production information in sufficient detail to enable a tender or tenders to be obtained.
    Application for statutory approvals.
F2 Preparation of further production information required under the building contract.

G   Tender Documentation
Preparation and collation of tender documentation in sufficient detail to enable a tender or tenders to be obtained for the construction of the project.

H   Tender Action
Identification and evaluation of potential contractors and/or specialists for the construction of the project.
Obtaining and appraising tenders and submission of recommendations to the client.

CONSTRUCTION PERIOD

J   Mobilisation
Letting the building contract, appointing the contractor.
Issuing of production information to the contractor.
Arranging site hand-over to the contractor.

K   Construction to Practical Completion
Administration of the building contract up to and including practical completion.
Provision to the contractor of further Information as and when reasonably required.

L   After Practical Completion
Administration of the building contract after practical completion.
Making final inspections and settling the final account.

Fig. 2.1

- the reason for the employer's desire to build – accommodation requirement, investment or profit
- whether or not the employer has a site and whether or not any preliminary discussions have taken place with the planning or other statutory authorities
- any preliminary work undertaken by the employer to establish a need or a market for the proposed project
- any technical skills that the employer can contribute
- any critical dates/timing.

From within their own organisations each member of the design team will need to assess:

- whether or not they are qualified to undertake the commission
- current workload – to help determine whether or not the commission is required and whether new staff will have to be recruited
- the level of risk – there may well be an element of project development work for which no fee will be paid if the project does not proceed beyond a certain stage, for example, the design team will often be expected to undertake a degree of speculative work in assisting with the preparation of a site bid, particularly when the prospective employer is a developer
- the effect of different procurement routes on their responsibilities and liabilities
- whether or not their professional indemnity insurance cover is adequate
- the employer's financial ability to complete the project.

The initial programme

One of the first questions the employer will ask will be: 'How quickly can I have my building?' At this very early stage it will be impossible for the design team to set a hard and fast programme. However, a simple bar chart (see Example 2/1) setting a timescale for the various key activities will provide both the employer and the design team with a useful framework within which to try to operate. The chart needs to separately identify any activities where the timescale is beyond the control of the design team and the employer, for example in obtaining development control and other statutory approvals.

At this stage the employer has to rely on the professional judgement of his advisers as to the likely implications of any difficulties. The chart contained within this book has assumed a straightforward project which should not encounter major delays.

In setting the initial programme it should be recognised that it is unlikely that the procurement route (be it drawings and bills of quantities, design and build or management contract) will have been established. Therefore, any

Month	SEPTEMBER				OCTOBER				NOVEMBER				DECEMBER				JANUARY				FEBRUARY				MARCH				APRIL				MAY				JUNE				JULY				AU	
Week Ending	1	8	15	22	29	6	13	20	27	3	10	17	24	1	8	15	22	29	5	12	19	26	2	9	16	23	2	9	16	23	30	6	13	20	27	4	11	18	25	1	8	15	22	29	6	13	20	27	3
Week Number	1	2	3	4	5	6	7	8	9	10	11	12	13	14	15	16	17	18	19	20	21	22	23	24	25	26	27	28	29	30	31	32	33	34	35	36	37	38	39	40	41	42	43	44	45	46	47	48	49

B - Stategic Briefing

C - Outline Proposals

D - Detailed Proposals

D - Development Control Application

E - Final Proposals

F - Production Information

F - Statutory Approvals

F - Sub-Contractors' Input

G - Tender Documentation

H - Tender Action

H - Report on Tenders & Let Contract

J - Mobilisation

K - Construction to Practical Completion

PROJECT BRIEF NOW FROZEN

DESIGN (FINAL PROPOSALS) NOW FROZEN - ANY
CHANGES WILL RESULT IN ABORTIVE WORK BY THE
DESIGN TEAM

COMPLETION OF PRODUCTION
INFORMATION AND SPECIFICATION
NOTES

DOCUMENTS SENT TO
TENDERING CONTRACTORS

TENDERS RETURNED

CONTRACT LET

WORKS START ON SITE

PROJECT PROGRAMME
VERSION 1.0

HEAVY LINES
INDICATE KEY
EVENTS

**Reed & Seymore
architects**

The Broadway : Borchester : BC4 2

Example 2/1   The initial programme.

The likely implications of any difficulties

assumptions which have been made in the absence of firm decisions should be confirmed during the development of the brief and adjustments made to the programme accordingly. The assessment of the resources required for the project and the agreement to an initial programme will assist the design team in advising on the development brief and the preparation of the detailed programme.

The appointment

Traditionally, an employer's first appointment to the design team is the architect, who will then assume the role of project leader. The architect will be expected to advise on the appointment of the other consultants to the team, although those appointments are invariably made direct with the employer. In this way the architect avoids contractual liability for the professional performance of the other members of the design team and for the payment of their fees. Increasingly, this traditional arrangement is being modified to suit alternative methods of building procurement which may result in a different project leader, for example a project manager (see Chapters 1 and 3).

Most employers invite fee bids from consultants for the particular service required. However, even when the employer provides the fullest possible information, when few details of a scheme are certain the preparation of realistic fee bids at an early stage is both difficult and financially hazardous. Equally, the sensible evaluation of competing bids is difficult. Just as with the submission of an unrealistically low building tender, the employer should be cautious of the unrealistically low fee bid and question the consultant's ability to satisfactorily perform the required services – although in any competitive tender/fee bid situation, it is unusual for anything other than the lowest price to be accepted.

Appointment documents

Some professional organisations make it compulsory for their members to provide written notification to their client of the terms and conditions of their appointment. Any member of the design team who belongs to an organisation that does not so require is advised to confirm, in writing to their client, at least the following:

- the name of the client
- the services that are to be provided
- the conditions of appointment
- how fees are to be calculated and when they are to be paid
- terms that will apply should the project change or not proceed.

Most professional organisations publish guides for clients relating to the appointment of their members together with standard terms of appointment and forms of agreement for execution by the client and design team member. Typically these documents will also provide for the specification of the services to be provided, often by way of a standard tick list, and the fee to be paid for carrying out those services.

Collateral warranties

In addition to the normal appointment documents consultants are frequently required to enter into collateral warranty agreements. Such agreements are used to create direct contractual relationships between any party other than the employer (e.g. funding institutions, purchasers, tenants) who is likely to suffer economic or consequential loss arising out of a construction project and any parties likely to cause such loss (e.g. consultants, contractor). The term collateral is used because the warranty is a side agreement to the warrantor's main contract with the employer.

In 1992 the British Property Federation Limited (BPF) published a series of standard Warranty Agreements including:

- CoWa/F – Collateral Warranty for funding institutions by a consultant
- CoWa/P&T – Collateral Warranty for purchasers and tenants by a consultant.

Both of these agreements, which are still used frequently, assume that the main contractor will be required to provide similar warranties. These are considered in Chapter 10.

The need for collateral warranties stemmed from the English law doctrine of privity of contract which provides that a person who is not a party to a contract cannot enforce that contract even if the obligations under the contract are for their benefit. On 11 May 2000 the Contracts (Rights of Third Parties) Act 1999 came into force and reformed the law relating to privity of contract by providing a person who is not a party to a contract with a right to enforce a contractual term if the contract expressly provides that they may do so, or if a term purports to confer a benefit on the third party by reference to a class of people. Whilst the Act is now a part of English law and will therefore apply to any contract governed by English law, it can be excluded from any contract by the inclusion of a simple contracting-out clause. JCT has, for the time being, opted to do this and in January 2000 introduced opting-out clause 1.12 into JCT 98.

Whilst this Act, in theory, removes the need for collateral warranty agreements, this has not as yet happened in the construction industry – partly because of potential problems associated with the determination of the extent of a party's liabilities and the securing of appropriate insurance to cover them and partly because of unfamiliarity with the new law.

Chapter 3
Procedure from Brief to Tender

Initial brief

The initial brief from the employer may be little more than a statement of intent, as mentioned in Chapter 2. However, for the project to be successful, the employer's brief must be developed in detail so that due consideration can be given to each of its aspects. The design brief and the design process develop in an iterative manner, with progress on one aspect creating a need for further thought on and consideration of other matters. The more complicated the building, the more important it is that the brief is developed in a systematic way and as early as possible. Full development, certainly on major schemes, is a team affair requiring the preparation of feasibility studies, cost reports and consultations with planning and other relevant authorities.

Developing the brief

Numerous factors need to be taken into account when developing the brief. Some of the main ones are:

- Employer attitude – establish whether the employer requires adherence to strict corporate standards.
- Environment – determine the type and quality of the internal and external environment desired.
- Operational factors – determine the activities the building is to accommodate and the other practical factors that will govern its layout and the relationship of the various elements within it. For example, assess the need to isolate certain items of plant and machinery in order to create acceptable noise and other environmental conditions elsewhere in the building. Also determine the extent of flexibility required in the designed space to provide for future adaptability.

- Site – identify ownership and any operational hazards. Evaluate access limitations, site topography and soil suitability (using soil tests, trial and bore holes). Survey the site and/or buildings to provide critical information such as access, boundaries, watercourses, trees, existing services, dimensions, levels and condition.
- Timescale – assess the appropriateness of the desired commencement and completion dates in the context of the practicalities of the situation and other projects that the employer may be planning and determine a programme.
- Finance – establish the employer's budget and funding arrangements and whether or not the envisaged scheme is likely to be varied if additional funds become available. Determine the relationship between capital and revenue funding and any cash flow constraints.
- Market – examine the market potential, timing and selling strategies for speculative buildings.
- Costs – assess overall projected capital and running costs.
- Development control – determine whether any particular development control considerations affect the site. Liaise as necessary with the development control authority.

Feasibility stage

Developing the design and costings will lead to a feasibility report, advising the employer whether the project is feasible functionally, technically and financially.

Sketch scheme

The design process usually commences with the sketch scheme. Initially this involves the designer putting on paper their preliminary responses to the brief. The sketch scheme will demonstrate the extent of accommodation that can be achieved: for example, the number of new houses on a particular site or the number of shopping units with their associated parking requirements. When the emphasis is on design quality as it would be in a conservation area or in the case of extensions to sensitive buildings, the sketch scheme will need also to address the visual form, style, proportion and materials. Whatever the nature of the drawings or information produced at this stage, the presentation should be capable of being readily understood by those reviewing the proposals.

Unless the employer has dealt with the issue, the design team should now advise on development control matters. They must determine whether or not the proposed development will require development control approval. If not, it is wise to obtain written confirmation of this in case of later dispute. Assuming approval is required, the need to obtain an outline development control approval prior to detail approval should be discussed. This usually becomes relevant when a principle needs to be established, for example a major change of use, the development of a site not designated under a local authority plan, or the development of a site over and above local density levels. Unless in a sensitive area, only basic plans or Ordnance Survey map extracts are required, thereby limiting the employer's financial commitment.

Costs

At the start of the sketch scheme the quantity surveyor will probably be required to prepare an initial cost estimate. This will invariably be based on a simple cost per square metre calculation and/or a comparison of outline unit costs. As the scheme develops it will be possible for the initial cost estimate to be refined on an element-by-element basis (see Chapter 4) such that the predicted cost may be amended or confirmed. This will establish whether or not the scheme is likely to be within the employer's budget. If it is not then appropriate adjustments can be made – these can be done more readily at this stage than later in the design process. Once the scheme is within budget the refined elemental estimate provides a cost control document. The cost of each element can then be monitored as the design develops and thereby prevented from unintentionally becoming unduly expensive or over-designed in comparison to other elements.

The employer is now in possession of an outline scheme which meets the basic parameters of his brief together with a cost plan based on known levels of quality. A decision as to whether or not to proceed can now be made. Beyond this point a commitment to progress the scheme will engender significant financial outlay.

Procurement

The method of choosing the contractor to construct the building must now be determined as this may dictate the manner in which the detailed design is progressed. For example, by following a traditional procurement route the design team will develop the design, whereas with design and build procurement the design team may only prepare a design brief, the design itself being completed by the contractor. Each procurement option will have a different time, cost and quality scenario (the procurement triangle). These three elements may be held in a particular balance as the circumstances dictate. However, one element must always give way to the others and altering any one element will have an effect on the others – see Fig. 3.1.

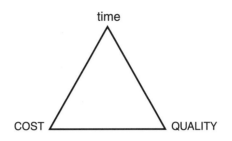

Fig. 3.1 The procurement triangle.

One interpretation of the time, cost and quality priorities which may be derived from three different procurement routes is given in Example 3/1. It must be stressed that the particular circumstances of each project may result in different conclusions as to the procurement route that best satisfies the priorities of time, cost and quality. For example, a high level of provisional sums within a contract could undermine the cost certainty otherwise afforded by

Procurement triangle

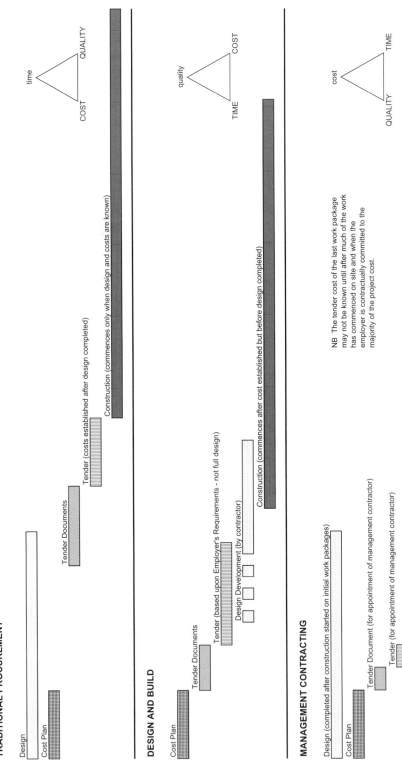

TRADITIONAL PROCUREMENT

Design

Cost Plan

Tender Documents

Tender (costs established after design completed)

Construction (commences only when design and costs are known)

time

COST — QUALITY

DESIGN AND BUILD

Cost Plan

Tender Documents

Tender (based upon Employer's Requirements - not full design)

Design Development (by contractor)

Construction (commences after cost established but before design completed)

quality

TIME — COST

MANAGEMENT CONTRACTING

Design (completed after construction started on initial work packages)

Cost Plan

Tender Document (for appointment of management contractor)

Tender (for appointment of management contractor)

Tender Documents (for work packages)

Construction (commences before design completed and cost established)

NB The tender cost of the last work package may not be known until after much of the work has commenced on site and when the employer is contractually committed to the majority of the project cost.

cost

QUALITY — TIME

Example 3/1   Procurement options.

a particular procurement route. Additionally, it should be noted that for ease of presentation time allocation for building control approval has been omitted from Example 3/1. This element in itself can influence the choice of the procurement route.

For more information on procurement routes and the forms of contract to use with them, refer to the Aqua Group's book *Tenders and Contracts for Building*.

It is advisable to apprise the employer of the procurement options, recommending the most appropriate route, detailing how the priorities are protected and outlining implications. Only when the method of procurement is determined can the pre-contract programme be finalised. This also represents the point in time when the specific services required of each member of the design team can be defined, fee proposals confirmed and agreements finalised.

Detail design

The development of the sketch scheme into a workable solution will produce the detail design to the employer's brief and allow it to be submitted for detailed development control approval. The onus is on the design team to maintain the priorities established within the employer's brief and to extract and agree any further information required from the employer as the scheme progresses. Whatever the scale of the project, the design team have a responsibility to use their skills to:

- produce a good solution to meet their employer's brief
- consider end and/or future users if different from the employer
- design to meet the anticipated lifespan of the building to avoid early deterioration or costly maintenance
- include, as far as reasonably practicable, health and safety considerations in the design
- co-ordinate structure, finishings and services into one complete design.

During this development stage it is wise for the designer to hold preliminary meetings with the local authority to assess the likelihood of development control approval, to define any agreements the employer may have to enter into with the authority and to determine the timescale involved. If reaction is unfavourable or objections from others seem likely, the designer owes a duty to the employer to advise him accordingly, indicating the possible extent of negotiation that may be required to achieve approval or the implications for time, cost and risk of refusal of appealing.

In submitting a detailed development control application the design team are likely to have to produce the following information:

- floor layout plans
- typical sections showing proposed heights
- all elevations
- elevations or pictorial views showing the proposals in context with adjoining buildings, where relevant
- site plans showing relationship of the proposal to other buildings, orientation, access, parking standards and the like
- details of materials and colours.

This information should be presented legibly and accurately in view of the different groups, committees, departments and members of the public that may have the right or duty to give their opinion.

If the application is refused or the application is not determined within an agreed timescale or the employer objects to certain conditions imposed in an approval, the applicant has the right to appeal within a time limit. Great care should be taken in advising an employer on these matters, especially in view of the penalties imposed for unreasonable or unsuccessful appeals. The employer may be well advised to seek expert legal advice prior to pursuing an appeal, particularly if the proposal is complex or risky. Once a detailed approval is granted an employer can instruct the design team to proceed to the production stage; if such instruction is given before the approval is granted the employer must accept the risk of abortive work.

Programming

When all submissions for statutory and other approvals have been satisfactorily completed and when the final design proposals have been given, the design team can re-examine the outline programme made at the start of the project and amend it in the light of the stage reached, the procurement option chosen and any financing conditions imposed.

The purpose of the detailed pre-contract element of the programme is to set out a sensible and logical sequence of the various pre-contract operations appropriate to the members of the design team and the required level of their input, together with external factors peculiar to the project. The initial bar chart or network can be expanded to show a programme to include:

- design team progress meetings
- preparing co-ordinated production information
- obtaining statutory approvals
- finalising legal agreements for access arrangements, party wall awards and the like
- negotiating with the service providers
- receiving and integrating specialist design elements

- receiving sub-contractors' and suppliers' quotations
- finalising information for the pre-tender health and safety plan
- preparing contract conditions and preliminaries including sequencing and contingencies
- preparing tender pricing documents
- preparing pre-tender estimate and updated cash flow prediction
- completion of pre-tender enquiries
- provision of tender documents
- receiving, appraising and reporting on tenders
- assembling contract documents
- briefing site inspectorate
- nominating/naming sub-contractors and suppliers.

In conjunction with these key points, the team should build in a contingency to cover such activities as co-ordination within the team, the inevitable drawing changes during the process and checking completed documentation. For the programme to run smoothly, all team members should be fully aware of the programme in respect of their own role, the interdependence of members of the design team and the effect that delay by one member will have on the others. It is the duty of the team to keep the employer informed of progress against the programme and to give advice on any factors which may alter the programme and their effect in cost and/or time.

Design team meetings

Chapter 2 refers to the traditional role of the architect as project leader. However, increasingly with larger projects the employer appoints a specific project manager to co-ordinate the pre-contract and site works on his behalf. The project leader should seek to manage the effective production of all the required information and ensure that the design team is working cohesively towards the common goal. He should be aware, therefore, of the terms of appointment and limits of responsibility of each team member and ensure that they work together to produce an efficient and coherent design.

It is easy for time to be wasted if the members of the design team are not in regular contact. The project leader should hold regular design team meetings to review:

- new and revised information
- the detailed design
- the integration of the structural and services elements
- specialist design input
- health and safety principles
- progress against the programme.

Clear decisions must be made at these meetings. Time spent constantly changing drawings is unproductive and likely to lead to abortive costs. A record
of decisions taken and any action required by a member of the design team
should be made so that such actions can be checked off at subsequent meetings.

It is important that the design team should not lose sight of the fact that
their role is to provide a service to the employer. Many professional practices
operate strict quality assurance procedures to ensure the maintenance of good
practice within their organisations and the provision of a quality service to the
employer.

Drawings

The most traditional and widespread method of conveying fundamental
information for building is the drawing. It is common for the architect to use
computer and overlay draughting techniques and to commence by producing
basic plans and sections (general arrangements) on which can be superimposed:

- setting-out dimensions
- structural engineer's designs
- services engineers' designs
- specialists' designs
- information for development control submissions.

Once setting-out, structure and services requirements have been determined,
the design team may then proceed to the production of large-scale details.

The clarity of drawings and other documents is fundamental to the smooth
running of any project. Hence the goal of the design team must be to make the
project information as simple and as easy to understand as possible. It is suggested that the most effective way of achieving this is to adopt the conventions
of Co-ordinated Project Information (CPI) as set out in a guide first published
by the Co-ordinating Committee for Project Information in 1987.

Specifications

In the early stages of the pre-contract programme the design team will need
to decide, in conjunction with the employer, appropriate performance requirements and/or the type and quality of all materials, goods and workmanship
necessary to complete the project. When selecting specific materials and goods
the design team will have to ensure that they:

- are appropriate to function, exposure and predicted use
- are available without undue difficulty
- provide appropriate levels of thermal resistance, sound attenuation and fire resistance
- comply with relevant health and safety conventions
- are environmentally friendly and energy efficient in manufacture and use
- create the required ambience in terms of space, colour and texture
- are easy to clean and maintain
- are appropriate to the employer's budget.

When translated into written form, the performance requirements and/or the qualitative details of the selected materials, goods and workmanship comprise the specification. In Chapter 6 we explain in more detail the different types of specification and their function.

Bills of quantities

The decisions on specification made by the design team and the employer form the basis of the written contract documentation. The method for presenting this information will depend upon the procurement route adopted. One of the more traditional procurement routes will require the production of bills of quantities. This process is one by which the quantity surveyor analyses the drawings and specification and, by following a standard set of rules, translates them into a schedule of quantified, descriptive items of the constituent parts of the proposed project.

The primary purpose of the bills of quantities is to provide a uniform basis for competitive lump sum tenders and a schedule of rates for pricing variations. Chapter 7 explains in more detail the different types of bills of quantities and their functions.

Specialist sub-contractors and suppliers

The use of certain materials, goods or installations requires specialist knowledge and skills that are likely to be beyond those possessed by the main contractor. When this occurs the design team will normally approach specialist suppliers or sub-contractors for the provision of design and quotations for the work. The methods by which this specialist information can be incorporated in the contract documents are discussed in Chapter 8.

Quality assurance

Throughout all procedures the quality of the product or service should be at an acceptable level. Quality assurance is a management process designed to provide a high probability that the defined objective of the product or service will be achieved. To achieve quality consistently, an organisation needs to have a management system in place. That system should be capable of applying appropriate management checks throughout all work stages, be it a design service, a product manufacturing process or a building erection process, and of correcting deficiencies if they occur.

Ground rules designed for manufacturing industry are laid down in BS 5750:1987 which, by careful interpretation, can be adjusted to suit professional services. The professional institutions have studied the British Standard, which mirrors ISO 9001, its international counterpart, and have published recommendations to their members on quality assurance.

A recognised way for a firm to set up its own quality assurance system is to prepare a quality manual covering:

- overall policy of the firm as regards quality of service
- policies on such matters as information services, staff training, resource control and documentation
- the preferred methods of running projects, such as those based on the Architect's Plan of Work (by RIBA Publications) or some other procedures manual.

The principals of the firm should formulate, endorse and evaluate the manual and communicate its contents to all personnel, ensuring full understanding. A firm may then decide whether or not it wishes to seek third-party assessment by a certification body leading to registration under BS 5750. Such a body will first examine whether the manual prepared by a firm meets all the requirements of the British Standard and then whether the firm is operating in accordance with the quality manual. After registration, further third-party assessments are usually carried out at regular intervals to ensure that the firm continues to operate in accordance with its quality manual.

Quality assurance on its own will not improve a firm's professional standards: it only provides an auditable record of performance against the firm's stated objectives, which may or may not embody high professional standards. The adoption of quality assurance will promote consistency in performance only at the level set down within the quality manual.

Obtaining tenders

In this chapter we have discussed the various elements involved in the pre-contract programme and factors important to the smooth running of that programme. We have also examined the different types of information that are used to convey to the contractor the work to be done and to obtain tender prices. In Chapters 4 to 8 these different types of information are discussed in greater detail and in Chapter 9 we recommend the procedure for obtaining selective tenders.

Chapter 4
Pre-Contract Cost Control

Cost control

To enable the employer to satisfactorily fund their desired building project they will need to know how much it is going to cost them and when they will have to pay. The service, by the design team, to provide this information is referred to in this book as cost control.

Ideally, cost control should be provided from inception to completion of the project such that the current estimated final cost is always known. Whilst we consider pre-contract cost control in terms of approximate estimates, cost plans, life cycle costing and cash flow predictions in this chapter, post-contract cost control is dealt with in Chapter 13 and capital allowances are dealt with in Chapter 17.

Reporting

If the design team are to obtain maximum benefit from pre-contract cost control it is essential that a sound cost reporting regime be adopted. The quantity surveyor should provide cost reports in accordance with the plan of work, (see Example 4/1). Also, because the level of information available on which to base the estimated costs will vary from time to time, it is essential that the quantity surveyor sets down in those reports, in clear and concise terms, as much of the following information as is available, stating where assumptions have been made:

- date of report
- names of the design team members
- address of the site
- extent of the site and any restrictions (e.g. as to use or access)
- planning situation
- basis of costs

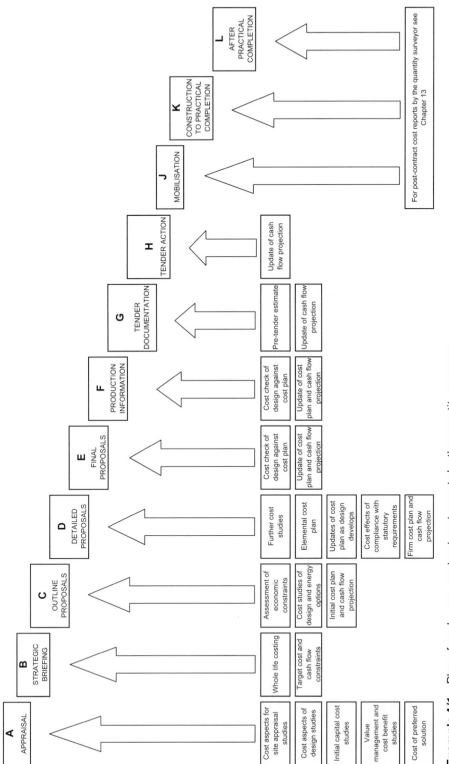

Example 4/1 Plan of work – pre-contract cost reports by the quantity surveyor.

- anticipated rate of inflation relating to costs used
- site investigation situation (e.g. site survey, ground investigation)
- requirement for any enabling works (e.g. demolitions, ground improvement)
- method of procurement
- date of commencement of the works on site
- contract period
- any phasing of possession
- any sectional completion
- schedule of drawings used
- outline of specification used
- availability of supply services (e.g. sewers, water, electricity, gas, telephone)
- exclusions (e.g. fitting out, professional and other fees, VAT)
- items with long delivery periods necessitating advance ordering
- a programme.

Establishing a budget

The detailed procedures to be adopted in pre-contract cost control will be governed, amongst other things, by the way in which the budget is established. There are three primary ways in which it is usual for this to be done:

(1)   Prescribed cost limit – the limit of capital expenditure is set down in the employer's brief, the amount normally being based on the maximum amount that can be afforded. 'Once in a lifetime' employers, not building for profit, will normally adopt this method of establishing a budget.

(2)   Unit cost limit – the limit of capital expenditure is calculated by the application of standard costs on a unit basis (e.g. £x per person to be accommodated, £y/m^2 of gross floor area to be provided). Public bodies and other employers who frequently procure buildings of a repetitive functional nature normally adopt this method of establishing a budget.

(3)   Determined cost limit – the limit of capital expenditure is based upon the economic viability of the proposed development. The employer will calculate how much of the non-recurring development costs are to be allocated to building costs if the scheme is to be economically viable. There is some latitude in this calculation, as the standard of building will, in part, determine the income to be derived from the scheme. All those building for profit will normally adopt this method of establishing a budget.

Approximate estimates

The object of the exercise is to estimate the cost to the employer of their desired scheme. It is usual to base the pricing of such estimates on historical cost data gleaned from the analyses of the costs of similar buildings adjusted to account for any fluctuations in price levels and for project idiosyncrasies.

The accuracy of the estimate will be influenced by the state of the design information, both drawings and specification notes. A **preliminary estimate** is often required when there is little or no information available, so it is usual to base such an estimate on place costs or net area/volume costs. As the design information is developed it is possible to produce more refined estimates based upon **element unit quantities** or measured **approximate quantities**. Finally, when the design is complete, an even more accurate estimate can be achieved by basing it on the measured **accurate quantities** in the bills of quantities.

Preliminary estimate

As the whole economics of a project are frequently based upon this estimate, it is not surprising that, of all the pre-contract cost information provided to the employer, the figures given at this time are the ones that will remain most clearly in his mind even though, by their very nature, they are the least accurate. It may well be that the preliminary estimate is required before any drawings, even simple sketch plans, have been prepared. Whether or not this is the case, it should not be produced until information on the following is to hand:

- the site and its nature
- the type of building and its use
- the overall height, number of floors and total floor area of each building
- an indication of the quality of materials and workmanship to be specified
- an outline of the engineering services to be provided.

The preliminary estimate will be based on either the units of accommodation or the net floor area/volume to be provided priced at appropriate £/unit, £/m² or £/m³ rates.

After completion of the preliminary estimate it will probably be necessary to carry out comparative cost studies of such matters as alternative plan shapes, numbers of floors, types of structure and other matters fundamental to the design.

Element unit quantities estimate

As more design information becomes available such an estimate may be prepared to check the accuracy of the preliminary estimate. The proposed buildings will be analysed in terms of element unit quantities (see the checklist of elements given later in this chapter) to which elemental unit rates are applied. The estimated cost of the external works and services will be based upon individually priced approximate quantities (see below). An element unit quantities estimate should not be produced until the following are available:

- Sketch plans:
  - 1:200 plan of each floor
  - 1:200 elevations
  - 1:200 sections
  - 1:500 or 1:200 site plan showing extent of external works and location of external services.
- Brief notes relating to the extent and/or standard of:
  - substructure
  - frame
  - upper floors
  - roof
  - stairs
  - external walls
  - windows and external doors
  - internal walls and partitions
  - internal doors
  - wall, floor and ceiling finishes
  - fittings and furnishings
  - internal services (e.g. sanitary fittings, disposal, water, heating, electrical, gas, lift, communications)
  - external services (e.g. disposal, water, electric, gas, telephone)
  - external works (e.g. roads, pavings, walls, fences, gates, water features).
- Brief notes on any unusual conditions of work or contract.

Approximate quantities estimate

This is usually produced when the proposed construction is unusual and/or complex or the project involves the alteration of existing building(s). This type of estimate is prepared on the basis of individually priced approximate quantities that are inclusive of all labours. Only the main items of work need be measured. Wherever possible items having similar measurements, for example hardcore bed, blinding, damp proof membrane, insulation, reinforcement and concrete slab, can be grouped together. Relatively low-cost complicated elements of work can be itemised and individually priced. In many instances a building can be analysed into as few as 100 measured items representing something of the order of 90% of the total cost. Provided that care is taken in booking the measurements and in choosing the rates, the approximate quantities estimate should be reasonably accurate. The information required is the same as that for the element unit quantities estimate plus:

- 1:100 plans of each floor
- 1:100 elevations of each face
- 1:100 sections
- typical details of important features.

Accurate quantities estimate

Such estimates are produced only when there are bills of quantities. These are priced before tenders are received to check the accuracy of previous estimates, to give advance warning of where adjustments in design may be necessary to reduce costs and to provide a base against which to compare the lowest tender.

Cost plan

When the design team are satisfied that they have determined the manner in which the employer's requirements can be best met, sketch plans and approximate estimates can be produced for the whole project and, as a prelude to the preparation of working drawings, a cost plan can be set up if this has not been done in conjunction with the earlier cost studies.

The cost plan is the document that brings design and cost together at pre-tender stage. When properly prepared and used it can assist the design team in both controlling the total building cost and spreading that cost between the various elements of the building in the most efficient manner, thus enabling them to make the best use of the employer's money.

In order to prepare a cost plan it is necessary to divide the building into a series of elements (e.g. frame, roof, upper floors). The choice of elements is determined by the nature of the building and the services and features in it. However, in order to maintain consistency in the compilation of cost data it is essential to always use the same basic elements. Perhaps the best way of achieving this is to adopt the listing of elements that has been used for many years by the Building Cost Information Service. This is reproduced below and provides a comprehensive checklist of elements and their sub-divisions.

(1) Substructure

All work below underside of screed or where no screed exists to underside of lowest floor finish including damp-proof membrane, together with relevant excavations and foundations.

...the best use of the employer's money

(2)   Superstructure

2.A   Frame

Loadbearing framework of concrete, steel or timber.

Main floor and roof beams, ties and roof trusses of framed buildings.

Casing to stanchions and beams for structural or protective purposes.

2.B   Upper floors

Upper floors, continuous access floors, balconies and structural screeds (access and private balconies each stated separately), suspended floors over or in basements stated separately.

2.C   Roof

2.C.1   Roof structure

Construction, including eaves and verges, plates and ceiling joists, gable ends, internal walls and chimneys above plate level, parapet walls and balustrades.

2.C.2   Roof coverings

Roof screeds and finishings.

Battening, felt, slating, tiling and the like.

Flashings and trims.

Insulation.

Eaves and verge treatment.

2.C.3   Roof drainage

Gutters where not integral with roof structure, rainwater heads and roof outlets. (Rainwater downpipes to be included in 'Internal drainage' (5.C.1).)

2.C.4   Rooflights
Rooflights, opening gear, frame, kerbs and glazing.
Pavement lights.

2.D   Stairs
2.D.1   Stair structure
Construction of ramps, stairs and landings other than at floor levels.
Ladders.
Escape staircases.

2.D.2   Stair finishes
Finishes to treads, risers, landings (other than at floor levels), ramp surfaces, strings and soffits.

2.D.3   Stair balustrades and handrails
Balustrades and handrails to stairs, landings and stairwells.

2.E   External walls
External enclosing walls including those to basements but excluding items included with 'Roof structure' (2.C.1).
Chimneys forming part of external walls up to plate level.
Curtain walling, sheeting rails and cladding.
Vertical tanking.
Insulation.
Applied external finishes.

2.F   Windows and external doors
2.F.1   Windows
Sashes, frames, linings and trim.
Ironmongery and glazing.
Shop fronts.
Lintels, sills, cavity damp-proof courses and work to reveals of openings.

2.F.2   External doors
Doors, fanlights and sidelights.
Frames, linings and trims.
Ironmongery and glazing.
Lintels, thresholds, cavity damp-proof courses and work to reveals of openings.

2.G   Internal walls and partitions
Internal walls, partitions and insulation.
Chimneys forming part of internal walls up to plate level.
Screens, borrowed lights and glazing.
Moveable space-dividing partitions.
Internal balustrades excluding items included with 'Stair balustrades and handrails' (2.D.3).

2.H   Internal doors
Doors, fanlights and sidelights.
Sliding and folding doors.

Hatches.
Frames, linings and trims.
Ironmongery and glazing.
Lintels, thresholds and work to reveals of openings.

(3) *Internal finishes*

3.A   Wall finishes
Preparatory work and finishes to surfaces of walls internally.
Picture, dado and similar rails.

3.B   Floor finishes
Preparatory work, screed, skirtings and finishes to floor surfaces
excluding items included with 'Stair finishes' (2.D.2) and structural
screeds included with 'Upper floors' (2.B).

3.C   Ceiling finishes
3.C.1   Finishes to ceilings
Preparatory work and finishes to surface of soffits excluding items
included with 'Stair finishes' (2.D.2) but including sides and soffits of
beams not forming part of a wall surface.
Cornices, coves.
3.C.2   Suspended ceilings
Construction and finishes of suspended ceilings.

(4) *Fittings and furnishings*

4.A   Fittings and furnishings
4.A.1   Fittings, fixtures and furniture
Fixed and loose fittings and furniture including shelving, cupboards,
wardrobes, benches, seating, counters and the like.
Blinds, blind boxes, curtain tracks and pelmets.
Blackboards, pin-up boards, notice boards, signs, lettering, mirrors and
the like.
Ironmongery.
4.A.2   Soft furnishings
Curtains, loose carpets or similar soft furnishing materials.
4.A.3   Works of art
Works of art if not included in a finishes element or elsewhere.
4.A.4   Equipment
Non-mechanical and non-electrical equipment related to the function
or need of the building (e.g. gymnasia equipment).

(5) *Services*

5.A   Sanitary appliances
Baths, basins, sinks, etc.

WCs, slop sinks, urinals and the like.

Toilet-roll holders, towel rails, etc.

Traps, waste fittings, overflows and taps as appropriate.

5.B   Services equipment

Kitchen, laundry, hospital and dental equipment and other specialist mechanical and electrical equipment related to the function of the building.

5.C   Disposal installations

5.C.1   Internal drainage

Waste pipes to 'Sanitary appliances' (5.A) and 'Services equipment' (5.B).

Soil, anti-siphonage and ventilation pipes.

Rainwater downpipes.

Floor channels and gratings and drains in ground within buildings up to external face of external walls.

5.C.2   Refuse disposal

Refuse ducts, waste disposal (grinding) units, chutes and bins.

Local incinerators and flues thereto.

Paper shredders and incinerators.

5.D   Water installations

5.D.1   Mains supply

Incoming water main from external face of external wall at point of entry into building including valves, water meters, rising main to (but excluding) storage tanks and main taps.

Insulation.

5.D.2   Cold water service

Storage tanks, pumps, pressure boosters, distribution pipework to sanitary appliances and to services equipment.

Valves and tanks not included with 'Sanitary appliances' (5.A) and/or 'Services equipment' (5.B).

Insulation.

5.D.3   Hot water service

Hot water and/or mixed water services.

Storage cylinders, pumps, calorifiers, instantaneous water heaters, distribution pipework to sanitary appliances and services equipment.

Valves and taps not included with 'Sanitary appliances' (5.A) and/or 'Services equipment' (5.B).

Insulation.

5.D.4   Steam and condensate

Steam distribution and condensate return pipework to and from services equipment within the building including all valves, fittings, etc.

Insulation.

5.E   Heat source

Boilers, mounting, firing equipment, pressurising equipment, instrumentation and control. ID and FD fans, gantries, flues and chimneys, fuel conveyors and calorifiers.

Cold and treated water supplies and tanks, fuel oil and/or gas supplies, storage tanks, etc., pipework (water or steam mains), pumps, valves and other equipment.

Insulation.

5.F   Space heating and air treatment

5.F.1   Water and/or steam

Heat emission units (radiators, pipe coils, etc.), valves and fittings, instrumentation and control and distribution pipework from 'Heat source' (5.E).

5.F.2   Ducted warm air

Ductwork, grilles, fans, filters, etc., instrumentation and control.

5.F.3   Electricity

Cable heating systems, off-peak heating systems, including storage radiators.

5.F.4   Local heating

Fireplaces (except flues), radiant heaters, small electrical or gas appliances, etc.

5.F.5   Other heating systems

Air treatment:

Air treated locally.

Air treated centrally.

Combination of treatments and whether inlet, extract or re-circulation.

High velocity system.

5.F.6   Heating with ventilation (air treated locally)

Distribution pipework, ducting, grilles, heat emission units including heating calorifiers except those which are part of 'Heat source' (5.E), instrumentation and control.

5.F.7   Heating with ventilation (air treated centrally)

All work as detailed under (5.F.6) for system where air treated centrally.

5.F.8   Heating with cooling (air treated locally)

All work as detailed under (5.F.6) including chilled water systems and/or cold treated water feeds. The whole of the costs of the cooling plant and distribution pipework to local cooling units shall be shown separately.

5.F.9   Heating with cooling (air treated centrally)

All work detailed under (5.F.8) for system where air treated centrally.

5.G   Ventilating system

Mechanical ventilating system not incorporating heating or cooling installations including dust and fume extraction and fresh air injection, unit extract fans, rotating ventilators and instrumentation and controls.

5.H Electrical installations

5.H.1   Electric source and mains

All work from external face of building up to and including local distribution boards including main switchgear, main and sub-main cables, control gear, power factor correction equipment, stand-by equipment, earthing, etc.

5.H.2   Electric power supplies

All wiring, cables, conduits, switches, etc., from local distribution boards, etc., to and including outlet points for the following:

- general-purpose socket outlets
- services equipment
- disposal installations
- water installations
- heat source
- space heating and air treatment
- gas installation
- lift and conveyor installations
- protective installations
- communication installations
- special installations.

5.H.3   Electric lighting

All wiring, cables, conduits, switches, etc. from local distribution boards and fittings to and including outlet points.

5.H.4   Electric lighting fittings

Lighting fittings including fixing.

Where lighting fittings supplied direct by client, this should be stated.

5.I Gas installations

Town and natural gas services from meter or from point of entry where there is no individual meter; distribution pipework to appliances and equipment.

5.J Lift and conveyor installations

5.J.1   Lifts and hoists

The complete installation including gantries, trolleys, blocks, hooks and ropes, pendant controls and electrical work from and including isolator.

5.J.2   Escalators

As detailed under 5.J.1.

5.J.3   Conveyors

As detailed under 5.J.1.

5.K Protective installations

5.K.1   Sprinkler installations

The complete sprinkler installation and CO_2 extinguishing system including tanks, control mechanism, etc.

5.K.2   Fire-fighting installations

Hose-reels, hand extinguishers, fire blankets, water and sand buckets, foam inlets, dry risers (and wet risers where only serving fire-fighting equipment).

5.K.3   Lightning protection

The complete lightning protection installation from finials and conductor tapes to and including earthing.

5.L   Communication installations

The following installations shall be included:

Warning installation (fire and theft)

Burglar and security alarms.

Fire alarms.

Visual and audio installations

Door signals.

Timed signals.

Call signals.

Clocks.

Telephones.

Public address.

Radio.

Television.

Pneumatic message system.

5.M   Special installations

All other mechanical and/or electrical installations (separately identified) which have not been included elsewhere, e.g. chemical gases; medical gases; vacuum cleaning; window cleaning equipment and cradles; compressed air; treated water; refrigerated stores.

5.N   Builder's work in connection with services

Builder's work in connection with mechanical and electrical services (5.A to 5.M each separately identified).

5.O   Builder's profit and attendance on services

Builder's profit and attendance in connection with mechanical and electrical services (5.A to 5.M each separately identified).

(6)   External works

6.A   Site works

6.A.1   Site preparation

Clearance and demolitions.

Preparatory earth works to form new contours.

6.A.2   Surface treatment

The cost of the following items shall be stated separately if possible:

- roads and associated footways
- vehicle parks
- paths and paved areas
- playing fields
- playgrounds
- games courts
- retaining walls
- land drainage
- landscape work

6.A.3   Site enclosure and division

Gates and entrance.

Fencing, walling and hedges.

6.A.4   Fittings and furniture

Notice boards, flag poles, seats, signs.

6.B   Drainage

Surface water drainage.

Foul drainage.

Sewerage treatment.

6.C   External services

6.C.1   Water mains

Main from existing supply up to external face of building.

6.C.2   Fire mains

Main from existing supply up to external face of building; fire hydrants.

6.C.3   Heating mains

Main from existing supply or heat source up to external face of building.

6.C.4   Gas mains

Main from existing supply up to external face of building.

6.C.5   Electric mains

Main from existing supply up to external face of building.

6.C.6   Site lighting

Distribution, fittings and equipment

6.C.7   Other mains and services

Mains relating to other service installations (each shown separately).

6.C.8   Builder's work in connection with external services

Builder's work in connection with external mechanical and electrical services, e.g. pits, trenches, ducts, etc. (6.C.1 to 6.C.7 each separately identified).

6.C.9   Builder's profit and attendance on external mechanical and electrical services

(6.C.1 to 6.C.7 each separately identified.)

6.D   Minor building work

6.D.1   Ancillary buildings

Separate minor buildings such as sub-stations, bicycle stores, horticultural buildings and the like, inclusive of local engineering services.

6.D.2   Alterations to existing buildings

Alterations and minor additions, shoring, repair and maintenance to existing buildings.

(7)   Preliminaries

Each element will have a particular design and/or specification characteristic and, although elements can influence each other, each will have a high percentage of its cost determined by design factors and the selection of material related directly to that element. Each element can therefore be considered in isolation until design, specification and costs are acceptable before its influences on other elements need to be examined.

The accuracy of the first cost plan will depend upon when and how it is prepared. If it is compiled at a very early stage by pricing element unit quantities (e.g. y m^2 of external wall, x Nr internal doors) at rates taken from a similar, earlier project, it is likely to be less accurate than if it is based on an elemental analysis of a project-specific approximate quantities estimate prepared at a later stage in the design process.

Any number of alternative design and specification solutions can be considered for each element. The cost of each should be estimated and superimposed on the cost plan to enable the design team to see how the various solutions affect cost. As the design is developed and decisions are made the cost plan must be updated. Where the value of any element is altered significantly by design development decisions, corresponding, financially inverse adjustments will have to be made to other elements if the project estimated cost is to be maintained within the established budget. For this to happen it is essential that there is close co-operation by, and good communication between, all in the design team throughout the design process.

By the time tenders are due to be received, the total of the cost plan should be a prediction of the lowest tender amount and should be at or below the established budget. If the lowest tender received exceeds the budget, the cost plan can be used to assist the design team in determining where savings can be made most effectively. As soon as a tender is accepted the cost plan should be revised so as to coincide with the contract sum; it can then be used as the basis for post-contract cost control.

Life cycle costing

Although the appearance of a building is generally of great importance, functional factors can be of higher priority. A building may well be judged by the employer not only on aesthetics but also on the basis of the practicality of design, running costs, maintenance costs, its ability to be remodelled in order to allow changes in use and the ease with which it can be extended. The design team must therefore ensure that the building will satisfy the user's requirements economically, not only initially but also for the whole life of the building. This will require a study of the effect of choice of design and specification on the cost to the employer of the building in use.

Life cycle costing (or whole life costing) is a technique used to assist in the identification of the most economically efficient way of achieving the best value in terms of capital and revenue expenditure (cost in use) on a building, an element of a building or even a component within a building. The basic method is to use normal discounted cash flow techniques to evaluate capital expenditure and running costs. In this way the case for higher initial expenditure on a building, element or component can be evaluated against lower ongoing costs in use.

It is important when undertaking such studies to anticipate an appropriate, effective life for the building. There is little point in choosing expensive materials solely for their durability if they are likely to outlast the useful life of the building.

Cash flow

The capital cost of a building project will normally be funded from the employer's reserves or by loans from financial institutions. The employer will need to know the dates and amounts of payments in advance if he is to be able to safeguard against unnecessary loss or charge of interest. The employer could require forecasts of payments to various third parties, the most frequent of which are:

- landowner – for site purchase
- contractor – for building work
- statutory authorities – for mains services connections
- service providers – for electricity, gas and water
- local authorities – for planning and building control fees
- consultants – for professional services
- financial institutions – for insurances, interest and other funding charges.

See Example 4/2 for a very simple cash flow forecast for building work based on the cost plan at pre-tender stage. When the contract is let it will become necessary to amend the forecast so that it reflects the contract sum, the date of possession and the date for completion.

PROJECT NAME
for
EMPLOYER'S NAME

Prepared 6 September 2001

DATE	EVENT	CERTIFIED AMOUNT £	VAT @ 17.5% £	TOTAL PAYMENT £
07/01/02	Date of Possession			
07/02/02	Interim certificate 1 issued			
21/02/02	**Final date for payment of interim certificate 1**	**133,000**	**23,275**	**156,275**
07/03/02	Interim certificate 2 issued			
21/03/02	**Final date for payment of interim certificate 2**	**209,000**	**36,575**	**245,575**
08/04/02	Interim certificate 3 issued			
22/04/02	**Final date for payment of interim certificate 3**	**247,000**	**43,225**	**290,225**
07/05/02	Interim certificate 4 issued			
21/05/02	**Final date for payment of interim certificate 4**	**209,000**	**36,575**	**245,575**
24/05/02	Date for Completion			
07/06/02	Interim certificate 5 issued			
21/06/02	**Final date for payment of interim certificate 5**	**99,000**	**17,325**	**116,325**
24/11/02	End of Defects Liability Period			
24/01/03	Final Certificate issued			
28/02/03	**Final date for payment of Final Certificate**	**23,000**	**4,025**	**27,025**
	TOTALS	**£920,000**	**£161,000**	**£1,081,000**

Example 4/2   Cash flow forecast for building work.

Chapter 5
Drawings and Schedules

Quality

Drawings represent the most important means by which the designer conveys his intention to the rest of the building team and to statutory authorities. As stated in Chapter 3, the clarity of drawings is fundamental to the smooth running of any project. This cannot be over-emphasised. Clear, concise, well-planned, co-ordinated drawings not only make the information they contain easy to understand; they also inspire confidence. Conversely, poor drawings do little except reveal the designer's lack of knowledge and inability to conduct affairs in a well-ordered manner.

The quality of information produced by the architect will either help or hinder his defence in any possible future legal action against him or for which he has to produce information for the defence of others. Recommendations on the preparation of drawings, especially working drawings, are contained in BS 1192:1984. Also important are the principles defined in BS 5750:1987 on quality assurance concerning the orderly preparation, dissemination and recording of design data and indeed all documentation relating to any form of manufacturing and production.

The method of producing drawings will vary from office to office depending on the size and resources of the practice. Some may rely heavily upon the use of computer-aided design, others on traditional draughting techniques, but most will rely on a combination of the two. The type of employer, the nature of the project, the degree of involvement with other consultants and their resources and, finally, the programme for the work all determine how the drawings at each stage are to be dealt with. But if quality is to be assured, there are simple rules that need to be observed no matter how the drawings are produced.

Quality procedures

An increasing number of firms have written a quality manual which is used to guide and control the quality and process of preparing documentation and conducting the professional and technical affairs of the firm. Quality manuals are assessed in action over a period of time by independent organisations before the latest international standard, BS EN ISO 9001:2000, can be awarded. This accreditation provides clients with a degree of assurance that the firm follows specific procedures in the way it deals with client assignments. An increasing number of clients in both the public and private sectors require that firms indicate they are quality assured when they make proposals for work. Quality manuals help firms to:

- provide an imaginative, competent and consistent professional service to their clients
- achieve, sustain and improve the quality of design services provided in a way that will consistently meet the stated, implied and perceived needs of clients
- provide the documented assurance to clients that the intended quality of service has been and will be achieved.

Quality procedures call for documents to be coded. An example is a four-part code comprising:

- **document code** to define the document type. These are:
  - MA – Manual
  - PR – Procedure
  - FM – Form
  - PN – Practice note
  - RR – Register (for system control)
- **responsibility code** to define the area of activity within the quality management system that is responsible for the writing, approving and distribution of system documents. These are:
  - AD – Administration
  - DM – Design management and project control
  - SM – System management and maintenance
  - MS – Management system information
- **document numbering code** to identify the serial number of the document within its own document type and responsibility code series. These can be simple two-digit numeric references such as 01.
- **status code** to identify the revision status of specific documents. It also distinguishes between documents that are still in draft format and those that have been brought into use. These are:

- OA – First draft
- OB – Second draft et seq.
- 01 – First issue
- 02 – Second issue et seq.

The procedures described below need to be established to review, audit, initiate corrective and preventive action, provide internal feedback and record client satisfaction (or dissatisfaction):

- **Management review.** Reviews of the quality system in the firm should be conducted by the quality manager at least twice yearly to evaluate its effectiveness and provide for continual improvement.
- **Internal audit.** All processes within the firm should be subject to systematic and independent internal audit by suitably trained staff in accordance with audit programmes.
- **Corrective and preventive action.** The cause of all non-conformity with procedures and standards is identified through:
  - risk assessment
  - audit
  - design review and verification
  - feedback
  - post-project review on selected projects.

The results should be analysed and appropriate corrective action implemented to prevent recurrence.

- **Internal feedback.** A simple mechanism should be established to enable all staff to propose changes and improvements to the quality system based on experience. All information and proposals should be considered and action taken recorded.
- **Client satisfaction (or dissatisfaction).** The level of client satisfaction should be recorded on both on-going and completed projects. This is effected through internal audit during the progress of projects and post-project review on completed projects.

Types, sizes and layout of drawings

The types of drawings vary as the project proceeds through its programme, as described later in this chapter. However, for efficiency and economy in the use of time, all drawings should:

- be on standard-sized sheets laid out in such a manner that the source and purpose of the drawing can be readily identified

- be shown to contain all necessary routine information and be able to be readily checked
- be kept in comprehensive sets and stored easily (record sets may have to be kept in archives for up to 25 years).

An assortment of drawings of different shapes and sizes with title panels and essential information in differing positions is a cause of confusion and irritation. Standardisation should apply no matter how the drawings are prepared.

In BS 3429:1984 recommendations are made for drawing sheet sizes A0, A1, A2, A3 and A4. In selecting the standard drawing sizes, account should be taken of the many and varied methods of reproduction: dye-line printing, photocopying, laser printing and photographic. Excluding routine dye-line printing, all methods available allow the possibility of changing the scale of the drawing in the course of reproduction – an invaluable asset in planning and producing comprehensive and co-ordinated sets of drawings. With this advantage in mind it is probable that the range of standard sheets might reasonably be limited to A0, A1 and A3.

A0 is a rather unwieldy size, especially for handling on site, but may be necessary for general arrangement drawings of large projects, townscape plans or extensive landscape drawings and the like. A1 size drawings are the most common and the most acceptable for ease of handling and can be reduced to any size down to A4. This is extremely useful in the preparation of design and presentation drawings that may be required in bound folders of A3 or A4 size. A3 sheets are most commonly used for quick sketches and details that may be reduced down to A4 size for use in transmission by fax.

With the almost universal use in offices of computers, the use of drawing sheets for the preparation of schedules is the exception rather than the rule. Standard letters, forms of instructions, certificates and most schedules are stored electronically for completion or amendment as the occasion arises. Drawing sheets might be used for window, door and ironmongery schedules which contain a considerable amount of information – in this case A3 sheets would be the most convenient.

The design of drawings must take account of both immediate and long-term storage. Whether it is for reduction and storage in files, in plan chests or in large clips hung on racks beside working stations, the layout of the sheet must provide sufficient margin to ensure that no information is obscured by the storage system.

Two types of title and information panel are recommended in BS 1192. The examples include space for classification references, but this will not be required unless a classified form of specification is being used. The latest and most widely adopted classification is the Unified Classification for the Construction Industry (Uniclass). This is a comprehensive coded system providing an interface between the manufacturer's product database, the use of materials and methods of assembly shown on the drawings, the specification and the bills of quantities.

Seldom are drawings produced and left unchanged. The development of the design and collaboration between members of the design team lead to a continuing process of amendment and revision. Every revision to a drawing must be adequately described and the date when it was made noted on the sheet and indicated by a unique revision code. If this is not done, it will be difficult, if not impossible, for other members of the building team to discover changes made to the drawings.

Most printing processes distort drawings – some are reduced in size arbitrarily for the convenience of storage or transmission by fax and some may be microfilmed for long-term storage of 'as built' information. It is therefore good practice to clearly incorporate the appropriate drawn scale(s) on every original drawing. In any event it must be remembered that it is always dangerous to take dimensions from any drawing by scaling.

Scale

Difficulties can arise if drawings are prepared to unusual scales. Engineering consultants frequently use the scales of 1:25 and 1:250. This can cause considerable confusion and should be discouraged. Another confusing scale is 'half size' as it is too easily mistaken for full size. It is recommended that the following scales be used:

...long term storage of 'as built' information

Location plans:
- 1:2500
- 1:1250
- 1:500

Site and development plans:
- 1:500
- 1:200

General arrangement drawings:
- 1:100
- 1:50

Component drawings:
- 1:50
- 1:20

Details:
- 1:5
- 1:1 (full size)

Assembly drawings:
- 1:20
- 1:10

Nature and sequence of drawing production

The type of contract and the method of choosing the contractor can affect the nature of the drawings and the sequence of their production. The building consists of many elements – structural frame, walls, partitions, roof, mechanical services and so on. Each of these elements forms a part of the whole and only a complete set of drawings can inform the contractor on the complete building. That does not necessarily mean that all drawings need to be completed before the contractor starts work on site – hence the need to understand, at the outset, the type of contract to be used.

The production of the drawings should be carefully planned. A preliminary list should be prepared of the drawings that will be required, not only from the architect but also from the other consultants.

Drawings for JCT98 contracts

When tenders are to be obtained and a contract placed on the basis of JCT98 with or without bills of quantities, it is predestined that all the drawings, schedules and specifications will be completed prior to obtaining tenders except when the contractor is provided with an information release schedule which states what information the architect will release and the time of its release. Reference to the Outline Plan of Work (see Chapter 2) will show how important it is to plan the production of the drawings, particularly through work stages E and F. It should also be borne in mind that where bills of quantities are used the SMM7R general rules set out details of the drawings required for the purposes of tendering.

The drawing production sequence is as follows:

- **General arrangement drawings.** Plans and elevations (1:100 or 1:50) which, after receiving the employer's approval, are sent to all consultants, who will then prepare their draft schemes.
- **Services drawings by the architect (plumbing, drainage).** These are worked out in detail concurrently with the preparation of the general arrangement drawings.
- **Construction details by the architect and specialist sub-contractors.** The selection of materials and finishes made during the preparation of the general arrangement drawings leads to the preparation of the architect's details and to obtaining tenders for specialist sub-contract design and construction. The outcome of this stage of the work can affect certain aspects, particularly critical dimensions in the primary elements of the structure and consequently the work of other consultants.
- **Assembly details.** When the consultants' drawings are accepted, the assembly details (1:20 and larger if necessary), which have been drafted in outline, should be completed.
- **Final co-ordination of drawings.** When all the detailed information has been assembled and co-ordinated, the final overall picture (originally the outline general arrangement drawings) can be completed with accuracy.
- **Layout and site plans.** These are finally completed, incorporating information on all external services and the setting out of the buildings.

Many details and drawings will be changed during the course of the works and, in adjusting the costs, it will be necessary to refer to the original information upon which the tender and contract sum was based. It is therefore of the utmost importance to keep a record set of the drawings used in the production of the other contract documents.

Drawings for design and build or management contracts

There are occasions when it is appropriate for the contractor to be responsible for the production of the building design or to manage a contract in such a manner as will accelerate the rate of construction (even though it might increase the cost). The appropriate forms of contract for such situations are considered in the Aqua Group's book *Tenders and Contracts for Building*. In such circumstances the preparation of the drawings will be geared not to the process of obtaining tenders for the whole of the works to be carried out by a single contractor but to obtaining sub-contractors' tenders for the various parts of the building, known as 'packages'. The tenders for these packages will be obtained in a sequence related to the programme of work to be carried out on site, on the basis of drawings and specifications and sometimes bills of quantities for each package. The order in which the drawings are produced will be related to the order of procurement and construction.

Once the building is designed, the sub-contract packages involved will be identified. The general arrangement drawings will have the same significance as in the traditional procurement route, but, once they are agreed, subsequent drawings will be of an elemental nature, that is, concerned with the various and separate elements of the building. Pile or foundation drawings, together with the main structural frame and drainage drawings, will be completely detailed and the work started on site, possibly before detailed or assembly drawings have started for secondary elements of the superstructure. Needless to say, the co-ordination of details in these circumstances becomes very difficult and can result in a rather different approach to design, calling for greater flexibility in the use of the structure or the building envelope to more readily allow substantial changes to the design late in its development.

The following is an example of the order in which drawings, specifications and bills of quantities (when used) for packages might be required. Those items marked with an asterisk (*) have a long lead-in or delivery period:

- general arrangement drawings
- levelling and general site works
- structural frame (steelwork)*
- drainage
- foundations (including piling)
- lifts and escalators*
- proprietary walling*
- roofing systems
- windows and doors*
- brickwork details
- mechanical and electrical engineering services*
- plant and machinery details
- partitions
- staircases and other secondary elements
- ceilings
- fixtures and fittings
- external works

There can be many more packages than are given here and some that are given separately may be combined.

Design intent information

In certain circumstances the design team may be asked to produce design intent drawings. One example context is that of a contractor working on a Private Finance Initiative (PFI) project as part of a consortium, who requires information to pass to his in-house design team.

The paradox of design intent drawings is that in many instances they are virtually complete as they have to communicate specific designs, details and specifications to another professional team and will be used for value management and detailed costing purposes. They will also be reviewed by the end user's team to assess value for money in the PFI context.

It is important that the design team led by the architect determines at the outset of the production information stage the exact level of detail required to be given in the design intent set.

Computer-aided design

The use of computers in architecture is now almost universal. The title given to this technique describes well the role of computers in the design process. Computers aid but they are not creative, rather they power-assist the creative process by enabling designers to assess their work accurately and rapidly. Computers are essential tools with many attractions to all the members of a design team. The main advantages lie in their ability to perform the technical services of producing drawings at great speed, of calculating and co-ordinating and ultimately of communicating. If there is a disadvantage, it may be seen in the type of drawings most commonly produced – of single line weight without enhancement or emphasis and often, for all their precision, lacking the special quality that allows easy reading and interpretation.

The most simple computer system will produce two-dimensional drawings containing no more information than is input for each drawing. The most advanced will store a complete model of the building and, on command, produce drawings to any scale or projection, of any component part, section or the whole of the building, while amendments to the original model will automatically appear on all drawings as they are produced. The storage capacity of even modestly powerful computers can carry a huge library of standard components, assemblies and details that will enable the architect or technician to produce quickly a vast range of working drawings of great accuracy.

Where members of the design team are using compatible hardware, discs can be exchanged giving each designer immediate access to all the information produced by the others. Layer upon layer, information can be added to the general arrangement drawings, ensuring complete co-ordination of all the parts of the building – structure, envelope details, mechanical and electrical engineering services and finishing details.

At the outset of the design process, members of the design team should discuss and agree the management of their CAD resources. All members of the team should produce one or more back-up copies of all drawings at all stages of production. Records of drawings undergoing major amendment should be stored. Common systems for coding drawings might well be adopted. On large projects, where many thousands of drawings may be produced, it is possible for the designer to have his system linked to specialist agencies which are made responsible for cataloguing and distributing to members of the building team the many drawings as they are produced or revised.

CAD enables every conceivable calculation related to the design process to be carried out, including structural, thermal performance, energy management, fire safety and lighting. And by following the procedures of CPI, the designers' programs can be interfaced with those of the quantity surveyor so that coded information on the drawings can be used in the production of the bills of quantities, specification documents and on-going cost estimates.

The use of CAD does not change the ground rules for good practice, but the systematic approach that it requires does help to improve the clarity of the information produced and makes it more accessible. Whichever way the drawings are produced, the information they convey should be the same.

Drawing file formats and translation

A wide range of CAD systems, with varying software and training costs, is now in use, responding to the needs of architecture, interior architecture and the various engineering disciplines. With this diversity of software and capability come associated problems of the quality of the interface between disparate systems. The .dxf format is one method of storing information for use across a wide range of software platforms. One simple and widely occurring example is exchanging data between Microstation and Autocad formats. However, this method is not infallible and, in the translation process, levels of detail may be lost.

At the job set-up stage of any project, the preferred CAD platform should be agreed, and if commonality cannot be achieved, a translation protocol should be devised. Design files should be checked carefully after translation to avoid errors of omission or distortion.

Project extranets

One of the more significant problems in the construction industry is that of the quality of communication between members of the building team. The traditional system is inherently inefficient. The employer is not part of the communication system and relies on maintaining a very close working relationship with the design team to keep up with their work. The method of transmission of information between members of the building team has not really changed in centuries.

Project extranets have been developed to overcome the wide range of problems related to the access of members of the building team to the most up-to-date information. There are at least 170 extranet companies, without commonality of format, although this volatile marketplace will, through failures and amalgamations, settle down to serve the market better.

Some systems are stronger during the design process, whilst some are stronger during construction. A project extranet is established along the lines of an intranet (using the Internet) to form an electronic project community. A file server holds all project information produced by all participants. This information can be accessed by members of the building team (including the employer) on an on-demand basis, from any location. Project information is classified according to stage, status and relevance. Access and the ability to manipulate information are carefully controlled. The user does not need to install a plethora of application software on his machines to view output from many sources.

Above all, a project extranet changes fundamentally the relationship between the members of the building team. From members being passive recipients of information received one from another, members become active participants, having to go to the extranet to extract information. This fundamental shift of responsibilities will improve the performance of the building team and ultimately the quality of service given to the employer.

Some extranets provide facilities for WebCams, so that site progress can be remotely checked by everyone concerned on a real-time basis, from anywhere in the world.

An important factor to consider when choosing a project extranet is to ensure that the chosen system will accommodate all the platforms used by the members of the building team. Most systems claim to run on browser technology but some operate only on Windows, thus excluding users of Unix, Macintosh and Linux. The need for a common 'engine' for all project extranets is important.

Project extranets provide the members of the building team with the ability to:

- share and organise project documents, plans and files
- access project information on a 24/7 basis in a secure, accessible, on-line location
- automate the creation and distribution of project updates and document revisions
- easily communicate with each other anywhere in the world
- monitor project status and progress
- access and view project audit trails and document version history

- red-line documents, issue, control and manage requests for information and add comments to or indicate changes on project drawings, contracts and other documents
- view on-line CAD output, photographs, spreadsheets, faxes and documents whilst maintaining data integrity
- automatically maintain relationships within compound documents
- set user-definable access control at the project, folder and document levels
- access a complete document history, including copies of earlier document versions, revisions and access records
- easily find specific documents within very large projects, via full text search and by document attributes such as file, name, type and author
- access disaster recovery facilities for project data
- access project-cloning capabilities, enabling all projects to be organised in the same way and allowing the rapid implementation of new projects.

The use of extranets allows business process management techniques to be deployed in the design and construction process, with the following advantages:

- shortens project schedules
- reduces costs
- increases productivity
- extends best practice tools and techniques throughout a project
- reduces project risks
- establishes full project accountability.

Extranet case study – BAA plc

BAA plc is the world's largest airport operator and has been using an extranet for part of its annual capital development programme, through the UK partner of a major US extranet provider. With a capital investment programme of around £1 million a day, BAA is one of the UK's principal developers of infrastructure and one of the construction industry's largest clients.

BAA and its framework partners are using a common extranet on a variety of projects, including new-build, extensions to existing buildings, refurbishments and maintenance. BAA has introduced new collaborative extranet technology across its Heathrow, Gatwick and Stansted airports to:

- improve efficiency and the timely sharing of the correct information
- reduce errors and wastage
- facilitate the re-use of information
- reduce life cycle costs
- improve the capturing of knowledge.

Contents of drawings

When planning and programming the production of drawings, account must be taken of the fact that every part of the building and its site has to be designed, detailed and drawn. Obvious as this might seem, it is unfortunately only too common to find elegant sets of drawings repetitiously showing the simple but assiduously avoiding the complicated. Should a part of a building be difficult to detail and to draw, that is the very detail and drawing that the contractor will need most. In considering how all the necessary information is to be conveyed, there are good and simple rules to observe:

- Draw every part of the building and do not repeat information unnecessarily.
- Sections are invaluable – indicate and code their location clearly.
- Remember that plans are horizontal sections; one plan for each floor level may not be enough.
- Number all rooms and spaces.
- Give reference numbers to all doors, windows, built-in fittings and the like; show these on all plans, sections and elevations.
- Make clear cross-reference to other drawings or to schedules so that detailed information can be traced.
- Be consistent in showing structural grids and levels on all plans, sections and elevations.
- Provide concise notes laid out clear of drawn information.
- Work out and indicate all essential dimensions but do not labour the obvious.

The following is a checklist of drawings and their contents.

Survey plan

- Existing site, boundaries and surrounds.
- Positions of major features such as existing buildings, roads, streams, ponds, walls and gates.
- Positions, girth, spread and type of trees.
- Location and type of hedgerows.
- Sufficient spot levels and contour lines, related to a specific datum, to enable a section of the site to be drawn in any direction.
- Position, invert and cover or surface levels of existing drains.
- Location, direction and depth of service mains.
- Rights of way, bridle paths and public footpaths.
- Access to site for vehicles.
- Ordnance reference if available.
- North point.
- A key plan showing the relationship of the various sheets, should the survey cover more than one sheet.

Site plan, layout and drainage

- Relevant information from the survey plan.
- Building profile and grid dimensionally related to a datum point or line.
- New roads and paths with widths and levels marked.
- Floor levels clearly indicated using the same datum as for the existing levels on the survey plan.
- Steps and ramps where they occur.
- Trees and hedgerows removed or retained.
- Street and other external lighting.
- Soil and surface water drains complete with pipe sizes and connections to sewers, making clear the distinction between soil and surface water drains and manholes (manhole sizes, levels and invert levels should be shown on a separate schedule).
- Gas, water and electric mains with depths indicated where possible and showing positions of connections to existing mains, supply companies' meters (external) and details of meter housings if required and points of termination within buildings.
- Banking and cutting and areas of disposing or spreading surplus soil.
- Details of fencing, existing and new.

General arrangement – for services

Once the design is approved and working drawings commenced, copy negatives should be taken of the general arrangement drawings of all floors showing only door swings in addition to walls and partitions and without any dimensions or other information on them. These drawings can then be developed without confusion or loss of time to show:

- Incoming mains and meter positions for all services.
- Electrical layout, including power points, light points, switches and all electrical equipment.
- Heating layout showing boilers, calorifiers, radiators or similar heating equipment.
- Plumbing and internal drainage layout.
- Gas layout.
- Sprinkler system layout, including pumps and storage tanks.
- Fire detection and protection equipment.
- Location of any special services.

General arrangement – foundations

- The main building grid annotated and dimensioned.

- Width and depth of all foundations for walls, piers and stanchions, with levels to underside.
- Positions and levels of drains, gullies and manholes close to foundations.
- Walls above foundations dotted with wall thicknesses dimensioned.
- Positions of incoming service mains, service main ducts and trenches and their levels.

General arrangement – floor plans

Complete plans drawn at a constant level through all openings at all floors and mezzanine floors:

- Overall dimensions.
- External dimensions, taking in all openings.
- Internal dimensions to show the positions and thicknesses of internal walls, partitions and all major features.
- Doors, their direction of swing and reference numbers.
- Windows and their reference numbers.
- Names or numbers of all rooms and circulation spaces (see BS 1192).
- Vertical and horizontal service ducts, holes in floor slabs, flues and builder's work in connection with services.
- Staircases, their direction of rise and stair treads numbered.
- Hatching to indicate materials of which walls and partitions are constructed.
- Rainwater pipes.
- Profile of joinery fittings and their code or reference numbers.
- Sanitary fittings.
- Floor levels clearly marked.

General arrangement – roof plan

- Main construction features.
- Levels clearly marked.
- Types of covering.
- Direction of falls.
- Rainwater outlets, gutters and pipes.
- Rooflights.
- Tank rooms, trap doors, chimneys, ventilation pipes and other penetrations; builder's work in connection with services.
- Parapets, copings and balustrades.
- Duckboards, catwalks, escape stairs and ladders.
- Lightning conductors, flag poles and aerials.

Elevations of all parts of the building

- New and old ground levels.
- Profiles of foundations.
- Damp-proof course levels.
- Levels of ground and upper floor slabs.
- External doors and windows with their reference numbers.
- Air bricks and ventilators.
- External materials, flashings.
- Rainwater pipes and gutters.
- Lightning conductors.

Sections that will describe all parts of the building and especially the relationship between separate but linked buildings

- New and original ground levels showing cut and fill.
- The main structural profile with grid lines.
- Type and depths of foundations.
- Floor and roof structures.
- Windows, doors and rooflights with their reference numbers.
- Lintels, arches and secondary structural members.
- Damp-proof membranes and courses, flashings.
- Eaves and valley gutters, parapets, rainwater heads and down pipes.
- All vertical dimensions.

Ceiling plans at all floor levels

- Room names and numbers over which ceilings occur.
- Type of ceiling, type of suspension, suspension pattern and finishes.
- Height of ceiling above floor level.
- Location of light fittings, their type and size.
- Fire detection and safety installations, ceiling void compartmentation, smoke detectors, emergency lighting.
- Mechanical ventilation registers.
- Sprinkler layouts.

Construction details (scale 1:20 and 1:10)

- Detailed sections for external walls, foundations and roofs.
- Plans, sections and elevations for staircases.
- Lift wells and escalators.

- Any room or part of the building the setting out of which is difficult, involves extensive fittings, fixtures, plumbing or special features or requires careful co-ordination, such as:
  - kitchens
  - bathrooms
  - lavatories
  - special-purpose rooms in hospitals, laboratories, etc.
- Building in window and door details.
- Part elevations and sections of the building's envelope containing special features such as:
  - entrances
  - special brick details
  - balconies
  - stonework
  - plant rooms.
- Builder's work in connection with services.
- Vertical and horizontal pipe ducts and their access.
- Fireplaces and flues.
- External details such as special paving, steps, kerbs, handrails and external lighting.

Large-scale details (scale 1:10 and 1:15)

Comprising enlargements of component parts of assemblies which have been shown to 1:20 scale but which require a larger scale to show the full details:

- Sills, heads and jambs of windows and doors.
- Special brickwork, string courses, arches (rubbed bricks) and special flashing details.
- Eaves, parapets, copings and special mouldings.
- Timber sections such as handrails, window sections and joinery details.
- Jointing details of curtain walling and other specialised cladding.
- Staircases.
- Special joinery fixtures and fittings.
- Any special feature which cannot be described or shown clearly to a smaller scale.

Schedules

It is good practice to convey information on items such as windows, doors and ironmongery by means of schedules which will set out the architect's requirements for other members of the building team in a manner which has many advantages over drawings. For example:

- Checking for errors of omission or duplication is simplified.
- Quantifying items for the purpose of obtaining estimates or placing orders is simplified.
- If the information is set down systematically, prolonged searching through a specification is avoided.

In setting down information in schedules, ensure clear layout, simple coding, the minimum of abbreviation and proper reference to the location in the building of the items concerned. Schedules should be either the same size as the standard drawing sheets selected for the job or, if smaller, bound together as a set in one of the accepted A sizes.

The design of and information contained in the title blocks should be consistent with those of the drawings. Schedules should also be numbered in a manner which is an extension of the numbering system used for the drawings. The first schedule should be a list of all the drawings and schedules related to the project, with space to record the latest revision number of each document. The first formal use of this schedule will be its inclusion in the specification or bills of quantities providing a record of the drawings and schedules used in the preparation of the tender documents.

The layout of schedules will vary according to the information to be conveyed. Items being scheduled may be listed to read downwards on the left-hand side, with the characteristics and sizes of the items reading across or vice versa. Diagrams may well be included in the body of the schedule and a column reserved for notes and records of revisions. Once an item has been fully described in a schedule it should not be repeated at length. For example, having described the construction of a particular type of door, it can be given a type reference and only this would be repeated in the schedule. Characteristics, particularly on a finishes schedule, can often be defined by a code with an interpretation in the form of a key to one side. Constant repetition of descriptions is then eliminated. Well-recognised abbreviations which will not be confused with others can be used to save time.

Schedules are not intended to supplant the specification or bills of quantities, which will contain the full and final description of the characteristics of the items concerned. Examples of schedules are given at the end of this chapter:

- Example 5/1 – Window schedule.
- Example 5/2 – Door schedule.
- Example 5/3 – Finishings schedule.
- Example 5/4 – Manhole schedule.

This is by no means a complete and comprehensive list – there are many other items which may be described and collated conveniently in schedules but it is very difficult for them to be standardised as they vary so much with each building. The examples given vary from those shown in BS 1192:1984 but they are more easily understood and less likely to lead to error.

Drawings and schedules for records

Although we have already referred in this chapter to the keeping of record drawings, it is a matter of sufficient importance to justify emphasis and clarification. Before starting work on site there are two stages at which it is important to keep record drawings and schedules before they are subjected to any further amendment:

- Documents upon which the specification or bills of quantities are based – a full set of completed drawings and schedules.
- Contract drawings and schedules – those to which the contract documents refer and which are marked accordingly.

It is useful for these copies to be kept in such form as permits them to be reproduced should the need arise.

True and accurate 'as built' drawings, including the exact routes of all services, are vital in the future maintenance of the building and copies should be made and issued to the employer upon completion of the contract as part of the health and safety file. Their preparation depends upon regular correction of the drawings as the work proceeds and as amendments are issued. They may be in plastic negative form so that they are easily reproduced and read or in microfilm form to minimise on storage space, although nowadays it is more likely that they will be stored and made available electronically.

WINDOWS

WINDOW TYPE REF:	A	B	C	D	E
WINDOW REF NO.	W2 to W19 inc. W22 to W39 inc.	W1, W20 W21, W40	W41, W42 and WW3	W44 to W50 inc.	W51
NUMBER OFF	36	4	3	7	1
MATERIAL OF MAIN FRAME	Extruded aluminium PVF2 treated, insulated sections: by Glazing International Limited				
SUB–FRAME	Nil	Nil	Nil	Nil	Hardwood
BUILDERS OPENING SIZE	Continuous horizontal opening height 1200		600Hx1500W	1800Hx750W	1800 diam
WINDOW UNIT SIZE	1185Hx1450W (inc cill)	1185Hx750W (inc. cill)	585Hx1485W	1785Hx735W	1485 diam
FIXING JAMB	To coupling mullion	Alum. brackets to brickwork	Aluminium brackets to brick		
HEAD	Alumium brackets plugged to concrete lintals or beam			Nil	
CILL	Aluminium cill with brackets plugged and screwed to blockwork			Nil	
SEALANT	Brown polysulphide to heads and jambs				

IRONMONGERY & GLAZING

FASTENINGS	Alum. lever handle and locking latch	Nil	Aluminium level handle	Aluminium peg stay	Nil
OPENING LIGHTS HINGES/PIVOTS	Horizontal hung projecting opening gear	Nil	Top hung (hinges with frame)	Side hung (hinges with frame)	Nil
STAYS	As part of opening gear	Nil	Pair of friction stays	Nil	Nil
OPERATING GEAR	Nil	Nil	Nil	Nil	Nil
SECURITY	Safety catch in jamb	Nil	Nil	Nil	Nil
GLAZING	Double glazed units, outer glass 6mm Pickertons Antison Silver, 12mm void, inner glass 6mm clear float				
FIXING	Extruded neoprene gaskets – all windows pre glazed by manufacturer				
NOTES					See drawing No. D02/64
REVISIONS Ref Wind Date A WA/21 28/2/92	DIAGRAMS				

Project Title   SHOPS & OFFICES - NEW BRIDGE STREET - BORCHESTER

Architects: REED & SEYMORE   WINDOW SCHEDULE   NO.   456/9.1   REV.   A

Example 5/1   Window schedule.

DOORS

DOOR TYPE REF:	A		B	C	D
DOOR REF NO.	1	2 to 8	1 to 4	1 to 8	1 to 15
LOCATION & NUMBER	Main entrance lobby 1	Entrance hall & corridor 01 & 11 7	Main staircase 4	Rooms 03, 04, 05 & 10 to 14 incl. 8	Offices 06 to 010 & 115 to 120 15
TYPE & DESCRIPTION	HW framed single swing double door; 6mm toughened glass		Plyfaced solid core flush panel with GWPP glazed viewing panel		Plyfaced solid core flush panel
FIRE RATING	Nil	Nil	60/60	30/30	Nil
DOOR SIZE	2 @ 2040x726x46 with 12mm rebated styles		2040x826x46 and 2040x426x46	2040x826x46	
FINISH	Polyurathene sealed		Colour preservative treated and wax polished		
FRAME OVERALL SIZE (nominal) & SECTION	2375x1850, 2375x1500 ex 150x60 rebated 25mm kicking rail ex 200x36		2375x1350 ex 150x50 rebated 25mm	2100x900 ex 150x60 rebated 25mm	2100x900 125x32 plus stops
MATERIAL/FINISH	Hardwood sealed		Softwood primed and painted		
FAN LIGHT	6mm PP		6mm GWPP	Nil	Nil
SIDE LIGHT	6mm toughened glass	Nil	6mm GWPP	Nil	Nil
SWING					
IRONMONGERY	1 @ 726 726	7 @ 726 726	4 @ 826 726	6 2	7 8
HINGES	1½ pair SS			1 pair rising butts	1 pair
LOCKS & LATCHES	Mortice deadlock type Ref. type –	Mortice dead lock type –	Mortice dead lock type –	Mortice latch and dead lock type –	
HANDLES	1 pair "D" pattern pull handles type –		Lever handles type –		
KICK PLATES	2 pairs satin aluminium 200 high		Nil	Nil	Nil
PUSH PLATES	1 pair satin aluminium 200 high		Nil	Nil	Nil
BOLTS	2 off SA flush (to leaf without lock)		2 off SA flush to small leaf	Nil	Nil
STOPS & STAYS		Skirting mounted type –		Floor mounted door stop type –	
MISCELLANEOUS			Surface mounted overhead closers. Intumescent strip to 3 sides		
REVISIONS Ref Door Date A DA/2 28/2/92	DIAGRAMS				

Project Title   SHOPS & OFFICES – NEW BRIDGE STREET – BORCHESTER

Architects: REED & SEYMORE   DOOR SCHEDULE   NO. 456/13.2   REV.   A

Example 5/2   Door schedule.

FINISHES

FLOOR LEVEL	Ground	Ground	Ground	First	First
ROOM NAME	Entrance Lobby	Stair Lobby	Enquiries	General Office	Womens Lavatory
ROOM NO.	G01	G02	G03	I01	I02

FINISHES

SCREED & FLOOR FINISH	75mm screed Carpet, Swatch ZD 10016			Raised floor system with carpet inlays to trays	50mm screed 150x150 ceramic floor tile
WALLS 1 / 2 / 3 / 4	Glazed Sirpite plaster finish to sand and cement base coat				Plaster Glazed Wall Tiles Glazed Wall Tiles Glazed Wall Tiles
SKIRTINGS	4x100x19 HW as detail 456/W/26			19x100 SW	Flush, coved ceramic tile
CEILING	Concealed grid suspension system; Class 1 tiles		Lay-in tile suspension system: Class 0/1		Plaster board, scrim & set
MISC & NOTES	Standard brass mat well frame set into screed	Staircase see detail drawing 456/14.3	Reception counter see drawing 456/W/25		

DECORATIONS

WALL FINISHES	Vinyl wall fabric by Sampsons Ltd	1 mist coat plus 2 coats vinyl emulsion – matt			Prime plus 2 coats flat oil
COLOURS 1 / 2 / 3 / 4	Glazed Ref No. SAM 1234 Ref No. SAM 1234 Ref No. SAM 1234	BS 00A01 BS 00A01 BS 00A01 BS 00A01	BS 20C37 BS 20C33 BS 20C33 BS 20C33	BS 16C33 BS 16C33 BS 16C33 BS 16C33	BS 18E51 2,3&4 tiled
CEILINGS	Nil	Nil	Nil	Nil	Flat oil white
FRAMES & ARCHITRAVES	1 + 2 coat gloss oil brilliant white				H2 gloss oil BS 20C40
DOORS	Glazed alumin	HW veneer 2 coats satin polyeurathene seal			Plastic laminate faced
WINDOWS	Aluminium frames self finished throughout				
CILLS	1 + 2 coats gloss oil brilliant white				Tiled as walls
SKIRTINGS	1 + 2 coats gloss oil brilliant white				1 + 2 gloss oil BS 20C40

REVISIONS Ref / Rm No. / Date	MISC & NOTES				
A G01 1/3/92		Handrails – 2 coats gloss polyeurathen seal	Reception counter – American oak french polished		Plastic laminate cubicles & ducting, extent of wall tiling etc, see drawing 456/W/27
B I02 16/3/92					

Project Title   SHOPS & OFFICES – NEW BRIDGE STREET – BORCHESTER

Architects: REED & SEYMORE   FINISHINGS SCHEDULE   NO. 456/14.1   REV. B

Example 5/3   Finishings schedule.

MANHOLES & COVERS

MANHOLE REF:	FM1	FM2	FM3	FM4	FM5
INTERNAL SIZE	1125x825	900 diam	1050 diam	1050 diam	1350 reducing to 900 diam
CONSTRUCTION & MAKE	225mm brick to 150 conc base	Precast concrete chamber rings, cover slabs and sealing rings set in concrete backfill. 150mm insitu concrete base			
COVER SIZE	450x600	450x600			
COVER TYPE	Med duty double-sealed	Grade C Fig 7		Grade C Fig 5	
INVERT LEVEL	7.315	7.214	7.02	6.807	5.705
GROUND LEVEL	8.46	8.153	8.175	8.08	8.07
COVER LEVEL	8.46	8.382	8.175	8.382	8.382
MAIN CHANNEL SIZE	100mm	100mm	100mm	100/150mm	150mm
TYPE STRAIGHT (S) CURVED (C) TAPPERED (T)	S	C	S	S T	S
BRANCH CHANNELS SIZE TO RH	2x100mmx90°		1x100mmx90° 1x100mmx135°	1x150mmx90°	
SIZES TO LH	1x100mmx135°	1x100mmx90°	1x100mmx135°	1x100mmx90° 1x100mmx135°	1x100mmx90°
INTERCEPTING TRAP	100mm				
F.A.I.					100mm
BACKDROP INLET					As detail drawing 071/C
STEP IRONS					C.I. into precast units
MISCELLANEOUS	DPC between head of brick and G.F slab		MH cover raised to paving level on brick upstand		
REVISIONS Ref MH No Date A FM4 28/2/92	DIAGRAMS Internal MH				

Project Title   SHOPS & OFFICES - NEW BRIDGE STREET - BORCHESTER

Architects: REED & SEYMORE   MANHOLES & COVER SCHEDULE   NO. 456/   REV. A

Example 5/4   Manhole schedule.

Chapter 6
Specifications

Introduction

The types and uses of specifications have changed dramatically in the past decade. They are no longer seen as documents that supplement the drawings or bills of quantities but as the key, stand-alone documents that define the scope and quality requirements in any construction contract.

For many years specifications comprised preambles to the bills of quantities, or brief written statements clarifying issues which could not be shown on the drawings. Occasionally they were used to describe a sequence of tasks to be performed. Today the picture is very different. Preambles have disappeared, drawings are produced using CAD technology and bills of quantities are often no more than a pricing schedule. The specification acts as the link between the matrices of contract documents, detailing not only scope, materials and workmanship requirements but also visual, procedural, design, testing, quality control and responsibility issues – the complete picture. The specification is now considered a prime document in any construction contract, second only in importance to the terms and conditions, for the simple reason that it records *the buy*.

In today's complex array of procurement, design and contractual methods, which seek to utilise the expertise of many contributors, the specification is the document that confirms precisely what one party agrees to provide to the other in respect of scope and quality. An inadequate clause or requirement, which conflicts with other contractual statements, will almost certainly lead to unwanted additional costs. The construction industry no longer relies on the 'design it, buy it then build it' methodology. Employer demands for high quality, speed of construction and cost control increasingly lead to the need for off-site fabrication where factory environments facilitate better quality control, economies of scale and less time on site. Traditional construction documentation is not always relevant to or understood by the manufacturing industry, and specifications therefore have to be adapted to suit. Modern buildings are very much a team effort, but legal responsibilities demand clear confirmations of who does what – the specification fulfils that requirement.

There are three basic types of specification:

- Prescriptive.
- Performance.
- Descriptive.

Prescriptive specifications

More commonly known as a detailed materials and workmanship specification, this is a 'do precisely as I say' specification. Such an approach pre-supposes that the design team know exactly what they want, have worked out exactly how it should be done, are allowed by European legislation to specify products, manufacturers, etc. and are prepared to warrant the design, taking full responsibility for the end product being fit for purpose. Prescriptive specifications are most commonly written for smaller projects using well-tried and tested technology.

Performance specifications

These are used in circumstances where the design team have absolutely no visual requirements but wish the contractor to select suitable materials and install

... for smaller projects using well-tried and tested technology

them to meet stated criteria. A performance specification does not confirm what will be provided, only what has to be achieved and as such the employer does not know what he is getting when entering into the contract – the employer's requirement is only that it performs and meets legislation and national standard requirements. A 'give me something that works' specification.

Descriptive specification

The 'new boys on the block' specification, this hybrid has been developed to make best use of the varying skills of those involved in the process of delivering modern buildings. The main function of a descriptive specification is to define scope, design intent, procedures for completing detailed design, quality control, and to provide the contractor with a fair indication of the solutions that are acceptable. The contractor is required to use his specialist expertise to complete the detailed design (in consultation with the design team), manufacture and install the works, and provide the necessary warranties and guarantees.

The need to prepare a contract specification which represents *the buy* is fundamental to the successful use of a descriptive specification if claims and variations are to be avoided at a later date. This requires a process of evaluating tenders in order to agree the key elements that reflect the scope of works. Descriptive specifications require that the design team and the contractor work in harmony and adopt a 'help me find the best solution' approach.

Large complex buildings tend to require the descriptive approach while the more traditional design solutions lend themselves to prescriptive documents.

Specification writing

Having established the various types and uses of specifications, there are a number of ways in which they can be prepared. Generally speaking, the following five processes are necessary in the production of a good specification:

- Decide on format.
- Collect information.
- Input information.
- Check and test.
- Deliver.

Decide on format

When selecting a format, three options are available:

- Uniclass.
- Masterformat.
- Bespoke.

Uniclass

The Unified Classification for the Construction Industry published by RIBA Publications is a classification scheme for organising library materials and for structuring product literature and project information. Section J, which comprises work sections for buildings, is reproduced at the end of this chapter with the permission of RIBA Enterprises Ltd. Uniclass is also used by the architect for the classification of drawings (see Chapter 5) and by the quantity surveyor, via SMM7R, in the preparation of bills of quantities (see Chapter 7). It therefore provides a useful link between the drawings, the bills of quantities (the pricing document) and the specification (the scope and quality document).

Care is required when using section JA because it mixes specification requirements for the permanent works with preliminaries requirements necessary to allow tenderers to price temporary, non-measurable and time-related items. It is important to remember that in the UK (unlike the USA) the specification is rarely the pricing document and bills of quantities no longer serve as the specification. Whilst the two should be cross-referenced for ease of use they should not be intermixed, although the specification may be incorporated in the bills of quantities in order to give it contract document status (see later in this chapter).

Another point worth noting is that the Uniclass work sections for buildings do not reflect construction trades or work packages. This results in most specifications being made up of several sections forming a rather bulky document. However, the proper use of sections JA7 (General specification for work packages) and JZ (Building fabric reference specification) helps to reduce repetition.

Masterformat

The Masterformat classification system is produced by the Construction Specifications Institute in the USA. It is the system that is recognised worldwide and is best adopted when working outside the UK and when working with large American firms in the UK. Masterformat is more construction trade related than Uniclass but it does include, as Section 1, General Condition, which is similar to our preliminaries. It is worth noting that bills of quantities are not generally recognised in the USA and that the specification is the document used there for pricing.

Bespoke

> Major employers commissioning multiple buildings occasionally produce specifications tailored to suit their particular needs, based on building type/size or a partnering procurement strategy. In most situations, however, the bespoke approach is not recommended as it is preferable to follow a recognised, published format.

Collect information

> It is to be remembered that the main purpose of any specification is to convey information from one party to another for contractual and control purposes. Accurate information is essential in order to provide *best information*. This can take many forms, ranging from the stipulation of a precise material and manufacturer to a detailed description of design intent. Both are equally effective provided that the specification format suits the procurement method and the selected form of contract. The wrong level of information put into a specification can lead to variations and claims for loss and expense.
>
> One good method of collection is to produce a list of materials or system descriptions, each with a unique short code such as 'BLK-1 = 100 mm common blockwork'. These codes can be used on the drawings and in the specification to provide clear scope and definition. This system is commonly known as technical sheet collection.
>
> The design team should decide at an early stage how best to gather information, bearing in mind the time available. Start collection early. Information should be updated continuously until the specification issue date. Continuous drafting and updating in parallel with drawing production is the best methodology. All changes should be recorded for quality control purposes.

Input information

> The key to accurate and fast completion of a specification is to use reliable and up-to-date source data. Off-the-shelf products are available. The best for use in the UK is perhaps the National Building Specification (NBS), which is arranged under the Uniclass work sections for buildings. It comprises detailed specification clauses together with skeleton clauses providing a sound checklist of the information necessary to properly specify preliminaries, general conditions, materials, goods and workmanship. NBS is updated regularly to take account of amendments to JCT Standard Forms of Building Contract and to reflect changes in technology and the latest standards for materials, goods and workmanship.

Masterspec is perhaps the best product for use in the USA when adopting Masterformat. Web-based products are also becoming available, such as davislangdonweb.net, which provides up-to-date information and a very fast production system. Whatever the source data, it is good practice for every clause in the specification to be uniquely numbered and for the project title, work section name and/or reference, version number, issue date and author's name to be given on every page. It is also good practice for drafts of the specification to be issued regularly for review and co-ordination by the design team.

Check and test

It is essential to ensure that there will be sufficient time available in which to test and check the specification before delivery. The author of a specification will never actually know how good it is until it is in use as a working document and is shown to provide clear direction when problems are encountered on site. However, it is a worthwhile exercise to test all specifications prior to their finalisation to ensure that they include the appropriate clauses to adequately resolve all problems previously encountered in work of a similar nature.

As for checking, it is almost impossible for the same person to be able to write and accurately proofread a document because the author rarely spots his own mistakes. It is therefore invariably beneficial to have a specification read through by a third party, who is preferably a competent specification writer and proofreader.

Deliver

There are two ways to deliver (issue) a specification:

- Hard copy.
- Electronic copy.

Hard copy

This involves numerous paper copies being delivered to all users and interested parties. These copies require close control and a record must be kept of exactly which version has been issued to whom, when and for what purpose. This method of delivery is expensive, slow, cumbersome and uses vast quantities of paper.

Electronic copy

The construction industry is slowly adapting to technology, and the use of electronic document management systems and web-based sites as a means of recording, storing and circulating information is on the increase. The use of CD-ROMs to circulate documents encounters similar problems to hard copy, except there is less bulk and less paper used. The use of e-mail is preferable, as it is fast, reliable and cheap. However, document identification and circulation control is just as important as it is with hard copy. Also, special consideration should be given to format – each recipient needs to be using the same software and settings in order to ensure that everyone is *seeing* the same document. It is strongly recommended that electronic files be issued always in an unexecutable format, such as PDF, in order to ensure that recipients do not make inappropriate alterations – when that happens it can be disastrous.

Web-based, dynamic production systems (e.g. davislangdonweb.net) are becoming available, providing faster, cheaper and better quality methods of production.

The specification as a contract document

Whilst JCT98 without quantities recognises the specification as a contract document, JCT98 with quantities does not. When using the latter form of contract, the specification can be given contract document status either by amending the wording of the standard form (not to be recommended) or by incorporating the specification into the bills of quantities, which are recognised as a contract document.

Work sections for buildings

J

JA	Preliminaries/General conditions

JA1 The project generally
JA10 Project particulars
JA11 Documentation
JA12 The site/Existing buildings
JA13 Description of the work

JA2 The Contract
JA20 The Contract/Sub-contract

JA3 Employer's requirements
JA30 Tendering/Sub-letting/Supply
JA31 Provision, content and use of documents
JA32 Management of the Works
JA33 Quality standards/control
JA34 Security/Safety/Protection
JA35 Specific limitations on method/sequence/timing/use of site
JA36 Facilities/Temporary works/Services
JA37 Operation/Maintenance of the finished building

JA4 Contractor's general cost items
JA40 Management and staff
JA41 Site accommodation
JA42 Services and facilities
JA43 Mechanical plant
JA44 Temporary works

JA5 Work by others or subject to instruction
JA50 Work/Materials by the employer
JA51 Nominated sub-contractors
JA52 Nominated suppliers
JA53 Work by statutory authorities
JA54 Provisional work
JA55 Dayworks

JA6 Preliminaries for specialist contracts
JA60 Demolition contract preliminaries
JA61 Ground investigation contract preliminaries
JA62 Piling contract preliminaries
JA63 Landscape contract preliminaries

JA7 General specification for work packages
JA70 General specification for building fabric work
JA71 General specification for building services work

JB	Complete buildings/ structures/units

JB1 Prefabricated buildings/ structures/units
JB10 Prefabricated buildings/structures
JB11 Prefabricated building units

JC	Existing site/buildings/ services

JC1 Investigations/Surveys
JC10 Site survey
JC11 Ground investigation
JC12 Underground services survey
JC13 Building fabric survey
JC14 Building services survey

JC2 Demolition/Removal
JC20 Demolition
JC21 Toxic/hazardous material removal

JC3 Alteration – support
JC30 Shoring/Facade retention

JC4 Repairing/Renovating/ Conserving concrete/masonry
JC40 Cleaning masonry/concrete
JC41 Repairing/Renovating/Conserving masonry
JC42 Repairing/Renovating/Conserving concrete
JC45 Damp proof course renewal/insertion

JC5 Repairing/Renovating/ Conserving metal/timber
JC50 Repairing/Renovating/Conserving metal
JC51 Repairing/Renovating/Conserving timber
JC52 Fungus/Beetle eradication

JC9 Alteration – composite items
JC90 Alterations – spot items

JD	Groundwork

JD1 Ground stabilisation/ dewatering
JD11 Soil stabilisation
JD12 Site dewatering

JD2 Excavation/filling
JD20 Excavating and filling
JD21 Landfill capping

JD3 Piling
JD30 Piling

JD4 Ground retention
JD40 Embedded retaining walls
JD41 Crib walls/Gabions/Reinforced earth

JD5 Underpinning
JD50 Underpinning

JE	In situ concrete/Large precast concrete

JE0 Concrete construction generally
JE05 In situ concrete construction generally

JE1 Mixing/Casting/Curing/ Spraying in situ concrete
JE10 Mixing/Casting/Curing in situ concrete
JE11 Sprayed concrete

JE2 Formwork
JE20 Formwork for in situ concrete

JE3 Reinforcement
JE30 Reinforcement for in situ concrete
JE31 Post tensioned reinforcement for in situ concrete

JE4 In situ concrete sundries
JE40 Designed joints in in situ concrete
JE41 Worked finishes/Cutting to in situ concrete
JE42 Accessories cast into in situ concrete

JE5 Structural precast concrete
JE50 Precast concrete frame structures

JE6 Composite construction
JE60 Precast/Composite concrete decking

Figure 6.1 *(Continued.)*

J   Work sections for buildings

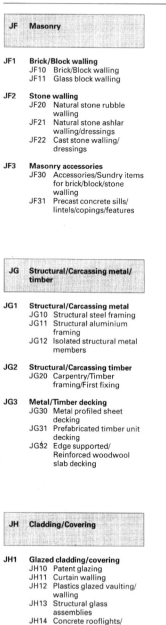

JF   Masonry

JF1   Brick/Block walling
JF10   Brick/Block walling
JF11   Glass block walling

JF2   Stone walling
JF20   Natural stone rubble walling
JF21   Natural stone ashlar walling/dressings
JF22   Cast stone walling/ dressings

JF3   Masonry accessories
JF30   Accessories/Sundry items for brick/block/stone walling
JF31   Precast concrete sills/ lintels/copings/features

JG   Structural/Carcassing metal/ timber

JG1   Structural/Carcassing metal
JG10   Structural steel framing
JG11   Structural aluminium framing
JG12   Isolated structural metal members

JG2   Structural/Carcassing timber
JG20   Carpentry/Timber framing/First fixing

JG3   Metal/Timber decking
JG30   Metal profiled sheet decking
JG31   Prefabricated timber unit decking
JG32   Edge supported/ Reinforced woodwool slab decking

JH   Cladding/Covering

JH1   Glazed cladding/covering
JH10   Patent glazing
JH11   Curtain walling
JH12   Plastics glazed vaulting/ walling
JH13   Structural glass assemblies
JH14   Concrete rooflights/ pavement lights
JH15   Rainscreen cladding/ overcladding

JH2   Sheet/board cladding
JH20   Rigid sheet cladding
JH21   Timber weatherboarding

JH3   Profiled/flat sheet cladding/ covering
JH30   Fibre cement profiled sheet cladding/covering
JH31   Metal profiled/flat sheet cladding/covering
JH32   Plastics profiled sheet cladding/covering
JH33   Bitumen and fibre profiled sheet cladding/ covering

JH4   Panel cladding
JH40   Glass reinforced cement panel cladding/features
JH41   Glass reinforced plastics panel cladding/features
JH42   Precast concrete panel cladding/features
JH43   Metal panel cladding/ features

JH5   Slab cladding
JH50   Precast concrete slab cladding/features
JH51   Natural stone slab cladding/features
JH52   Cast stone slab cladding/ features

JH6   Slate/Tile cladding/covering
JH60   Plain roof tiling
JH61   Fibre cement slating
JH62   Natural slating
JH63   Reconstructed stone slating/tiling
JH64   Timber shingling
JH65   Single lap roof tiling
JH66   Bituminous felt shingling

JH7   Malleable sheet coverings/cladding
JH70   Malleable metal sheet pre-bonded coverings/ claddings
JH71   Lead sheet coverings/ flashings
JH72   Aluminium sheet coverings/flashings
JH73   Copper strip/sheet coverings/flashings
JH74   Zinc strip/sheet coverings/flashings
JH75   Stainless steel strip/sheet coverings/flashings
JH76   Fibre bitumen thermoplastic sheet coverings/flashings

JH9   Other cladding/covering
JH90   Tensile fabric coverings
JH91   Thatch roofing

JJ   Waterproofing

JJ1   Cementitious coatings
JJ10   Specialist waterproof rendering

JJ2   Asphalt coatings
JJ20   Mastic asphalt tanking/ damp proofing
JJ21   Mastic asphalt roofing/ insulation/ finishes
JJ22   Proprietary roof decking with asphalt finish

JJ3   Liquid applied coatings
JJ30   Liquid applied tanking/ damp proofing
JJ31   Liquid applied waterproof roof coatings
JJ32   Sprayed vapour barriers
JJ33   In situ glass reinforced plastics

JJ4   Felt/flexible sheets
JJ40   Flexible sheet tanking/damp proofing
JJ41   Built-up felt roof coverings
JJ42   Single layer polymeric roof coverings
JJ43   Proprietary roof decking with felt finish
JJ44   Sheet linings for pools/ lakes/waterways

JK   Linings/Sheathing/Dry partitioning

JK1   Rigid sheet sheathing/linings
JK10   Plasterboard dry lining/ partitions/ceilings
       For panel partitions see JK30.
JK11   Rigid sheet flooring/ sheathing/linings/casings
JK12   Under purlin/Inside rail panel linings
JK13   Rigid sheet fine linings/ panelling
JK14   Glass reinforced gypsum linings/panelling/casings/ mouldings
JK15   Vitreous enamel linings/ panelling

JK2   Timber board/Strip linings
JK20   Timber board flooring/ sheathing/linings/casings
JK21   Timber strip/board fine flooring/linings

Figure 6.1   *(Continued.)*

Work sections for buildings J

JK3 Dry partitions
JK30 Panel partitions
JK31 *intentionally not used*
JK32 Framed panel cubicles
JK33 Concrete/Terrazzo partitions

JK4 False ceilings/floors
JK40 Demountable suspended ceilings
JK41 Raised access floors

JL Windows/Doors/Stairs

JL1 Windows/Rooflights/ Screens/Louvres
JL10 Windows
JL11 Rooflights/ Roof windows
JL12 Screens
JL13 Louvred ventilators
JL14 External louvres/ shutters/canopies/blinds

JL2 Doors/Shutters/Hatches
JL20 Doors
JL21 Shutters
JL22 Hatches

JL3 Stairs/Walkways/Balustrades
JL30 Stairs/Walkways/ Balustrades

JL4 Glazing
JL40 General glazing
JL41 Lead light glazing
JL42 Infill panels/sheets

JM Surface finishes

JM1 Screeds/Trowelled flooring
JM10 Cement:sand/Concrete screeds/toppings
JM11 Mastic asphalt flooring/ floor underlays
JM12 Trowelled bitumen/resin/ rubber-latex flooring
JM13 Calcium sulfate based screeds

JM2 Plastered coatings
JM20 Plastered/Rendered/ Roughcast coatings
JM21 Insulation with rendered finish
JM22 Sprayed monolithic coatings
JM23 Resin bound mineral coatings

JM3 Work related to plastered coatings
JM30 Metal mesh lathing/ Anchored reinforcement for plastered coatings
JM31 Fibrous plaster

JM4 Rigid tiles
JM40 Stone/Concrete/Quarry/ Ceramic tiling/Mosaic
JM41 Terrazzo tiling/In situ terrazzo
JM42 Wood block/Composition block/Parquet flooring

JM5 Flexible sheet/tile coverings
JM50 Rubber/Plastics/Cork/ Lino/Carpet tiling/ sheeting
JM51 Edge fixed carpeting
JM52 Decorative papers/fabrics

JM6 Painting
JM60 Painting/Clear finishing
JM61 Intumescent coatings for fire protection of steelwork

JN Furniture/Equipment

JN1 General purpose fixtures/ furnishings/equipment
JN10 General fixtures/ furnishings/equipment
JN11 Domestic kitchen fittings
JN12 Catering equipment
JN13 Sanitary appliances/ fittings
JN14 Plant containers
JN15 Signs/Notices

JN2 Special purpose fixtures/ furnishings/equipment
JN20 Appropriate section title for each project
JN21 Appropriate section title for each project
JN22 Appropriate section title for each project
JN23 Appropriate section title for each project

JP Building fabric sundries

JP1 Sundry proofing/insulation
JP10 Sundry insulation/ proofing work/fire stops
JP11 Foamed/Fibre/Bead cavity wall insulation

JP2 Sundry finishes/fittings
JP20 Unframed isolated trims/skirtings/sundry items
JP21 Ironmongery
JP22 Sealant joints

JP3 Sundry work in connection with engineering services
JP30 Trenches/Pipeways/Pits for buried engineering services
JP31 Holes/Chases/Covers/ Supports for services

JQ Paving/Planting/Fencing/Site furniture

JQ1 Edgings/Accessories for pavings
JQ10 Kerbs/Edgings/Channels/ Paving accessories

JQ2 Pavings
JQ20 Granular sub-bases to roads/pavings
JQ21 In situ concrete roads/pavings/bases
JQ22 Coated macadam/ Asphalt roads/pavings
JQ23 Gravel/Hoggin/Bark roads/pavings
JQ24 Interlocking brick/block roads/pavings
JQ25 Slab/Brick/Sett/Cobble pavings
JQ26 Special surfacings/ pavings for sport/general amenity

JQ3 Planting
JQ30 Seeding/Turfing
JQ31 Planting
JQ32 Planting in special environments
JQ35 Landscape maintenance

JQ4 Fencing
JQ40 Fencing

JQ5 Site furniture
JQ50 Site/Street furniture/ equipment

Figure 6.1 (*Continued.*)

J   Work sections for buildings

JR	Disposal systems

JR1 Drainage
- JR10 Rainwater pipework/gutters
- JR11 Foul drainage above ground
- JR12 Drainage below ground
- JR13 Land drainage
- JR14 Laboratory/Industrial waste drainage

JR2 Sewerage
- JR20 Sewage pumping
- JR21 Sewage treatment/sterilisation

JR3 Refuse disposal
- JR30 Centralised vacuum cleaning
- JR31 Refuse chutes
- JR32 Compactors/Macerators
- JR33 Incineration plant

JS	Piped supply systems

JS1 Water supply
- JS10 Cold water
- JS11 Hot water
- JS12 Hot and cold water (small scale)
- JS13 Pressurised water
- JS14 Irrigation
- JS15 Fountains/Water features

JS2 Treated on site water supply
- JS20 Treated/Deionised/Distilled water
- JS21 Swimming pool water treatment

JS3 Gas supply
- JS30 Compressed air
- JS31 Instrument air
- JS32 Natural gas
- JS33 Liquefied petroleum gas
- JS34 Medical/Laboratory gas

JS4 Petrol/Oil storage
- JS40 Petrol/Diesel storage/distribution
- JS41 Fuel oil storage/distribution

JS5 Other supply systems
- JS50 Vacuum
- JS51 Steam

JS6 Fire fighting – water
- JS60 Fire hose reels
- JS61 Dry risers
- JS62 Wet risers
- JS63 Sprinklers
- JS64 Deluge
- JS65 Fire hydrants

JS7 Fire fighting – gas/foam
- JS70 Gas fire fighting
- JS71 Foam fire fighting

JT	Mechanical heating/Cooling/Refrigeration systems

JT1 Heat source
- JT10 Gas/Oil fired boilers
- JT11 Coal fired boilers
- JT12 Electrode/Direct electric boilers
- JT13 Packaged steam generators
- JT14 Heat pumps
- JT15 Solar collectors
- JT16 Alternative fuel boilers

JT2 Primary heat distribution
- JT20 Primary heat distribution

JT3 Heat distribution/utilisation – water
- JT30 Medium temperature hot water heating
- JT31 Low temperature hot water heating
- JT32 Low temperature hot water heating (small scale)
- JT33 Steam heating

JT4 Heat distribution/utilisation – air
- JT40 Warm air heating
- JT41 Warm air heating (small scale)
- JT42 Local heating units

JT5 Heat recovery
- JT50 Heat recovery

JT6 Central refrigeration/Distribution
- JT60 Central refrigeration plant
- JT61 Chilled water

JT7 Local cooling/Refrigeration
- JT70 Local cooling units
- JT71 Cold rooms
- JT72 Ice pads

JU	Ventilation/Air conditioning systems

JU1 Ventilation/Fume extract
- JU10 General ventilation
- JU11 Toilet ventilation
- JU12 Kitchen ventilation
- JU13 Car parking ventilation
- JU14 Smoke extract/Smoke control
- JU15 Safety cabinet/Fume cupboard extract
- JU16 Fume extract
- JU17 Anaesthetic gas extract

JU2 Industrial extract
- JU20 Dust collection

JU3 Air conditioning – all air
- JU30 Low velocity air conditioning
- JU31 VAV air conditioning
- JU32 Dual-duct air conditioning
- JU33 Multi-zone air conditioning

JU4 Air conditioning – air/water
- JU40 Induction air conditioning
- JU41 Fan-coil air conditioning
- JU42 Terminal re-heat air conditioning
- JU43 Terminal heat pump air conditioning

JU5 Air conditioning – hybrid
- JU50 Hybrid system air conditioning

JU6 Air conditioning – local
- JU60 Air conditioning units

JU7 Other air systems
- JU70 Air curtains

JV	Electrical supply/power/lighting systems

JV1 Generation/Supply/HV distribution
- JV10 Electricity generation plant
- JV11 HV supply/distribution/public utility supply
- JV12 LV supply/public utility supply

JV2 General LV distribution/lighting/power
- JV20 LV distribution
- JV21 General lighting
- JV22 General LV power

JV3 Special types of supply/distribution
- JV30 Extra low voltage supply
- JV31 DC supply
- JV32 Uninterrupted power supply

Figure 6.1   (*Continued.*)

Work sections for buildings

J

JV4 Special lighting
JV40 Emergency lighting
JV41 Street/Area/Flood lighting
JV42 Studio/Auditorium/Arena
lighting

JV5 Electric heating
JV50 Electric underfloor/ceiling
heating
JV51 Local electric heating
units

JV9 General/Other electrical work
JV90 General lighting and
power (small scale)

**JW Communications/Security/
Control systems**

**JW1 Communications – speech/
audio**
JW10 Telecommunications
JW11 Paging/Emergency call
JW12 Public address/
Conference audio
facilities

JW2 Communications – audio-visual
JW20 Radio/TV/CCTV
JW21 Projection
JW22 Information/Advertising
display
JW23 Clocks

JW3 Communications – data
JW30 Data transmission

JW4 Security
JW40 Access control
JW41 Security detection and
alarm

JW5 Protection
JW50 Fire detection and alarm
JW51 Earthing and bonding
JW52 Lightning protection
JW53 Electromagnetic
screening
JW54 Liquid detection alarm
JW55 Gas detection alarm
JW56 Electronic bird/vermin
control

JW6 Central control
JW60 Central control/Building
management

JX Transport systems

JX1 People/Goods
JX10 Lifts
JX11 Escalators
JX12 Moving pavements
JX13 Powered stairlifts
JX14 Fire escape chutes/slings

JX2 Goods/Maintenance
JX20 Hoists
JX21 Cranes
JX22 Travelling cradles/
Gantries/Ladders
JX23 Goods distribution/
Mechanised warehousing

JX3 Documents
JX30 Mechanical document
conveying
JX31 Pneumatic document
conveying
JX32 Automatic document
filing and retrieval

**JY Services reference
specification**

JY1 Pipelines and ancillaries
JY10 Pipelines
JY11 Pipeline ancillaries

JY2 General pipeline equipment
JY20 Pumps
JY21 Water tanks/cisterns
JY22 Heat exchangers
JY23 Storage cylinders/
Calorifiers
JY24 Trace heating
JY25 Cleaning and chemical
treatment

JY3 Air ductlines and ancillaries
JY30 Air ductlines/ancillaries

JY4 General air ductline equipment
JY40 Air handling units
JY41 Fans
JY42 Air filtration
JY43 Heating/Cooling coils
JY44 Air treatment
JY45 Silencers/Acoustic
treatment
JY46 Grilles/Diffusers/Louvres

**JY5 Other common mechanical
items**
JY50 Thermal insulation
JY51 Testing and
commissioning of
mechanical services
JY52 Vibration isolation
mountings
JY53 Control components –
mechanical
JY54 Identification –
mechanical
JY59 Sundry common
mechanical items

JY6 Cables and wiring
JY60 Conduit and cable
trunking
JY61 HV/LV cables and wiring
JY62 Busbar trunking
JY63 Support components –
cables

JY7 General electrical equipment
JY70 HV switchgear
JY71 LV switchgear and
distribution boards
JY72 Contactors and starters
JY73 Luminaires and lamps
JY74 Accessories for electrical
services

JY8 Other common electrical items
JY80 Earthing and bonding
components
JY81 Testing and
commissioning of
electrical services
JY82 Identification – electrical
JY89 Sundry common
electrical items

**JY9 Other common mechanical
and/or electrical items**
JY90 Fixing to building fabric
JY91 Off-site painting/Anti-
corrosion treatments
JY92 Motor drives – electric

**JZ Building fabric reference
specification**

JZ1 Fabricating
JZ10 Purpose made joinery
JZ11 Purpose made metalwork
JZ12 Preservative/Fire
retardant treatments for
timber

JZ2 Fixing/Jointing
JZ20 Fixings/Adhesives
JZ21 Mortars
JZ22 Sealants

JZ3 Finishing
JZ30 Off-site painting
JZ31 Powder coatings
JZ32 Liquid coatings
JZ33 Anodising

Figure 6.1 *(Continued.)*

Chapter 7
Bills of Quantities

Tender and contract document

Bills of quantities are documents that describe the quality and give the quantities of the constituent parts of proposed building works. They have two primary functions. Initially they are used as tender procurement documents to provide a uniform basis for competitive lump sum tenders. Subsequently they become contract documents serving as schedules of rates for the pricing of variations. It is therefore important that they contain at least the basic information required of them by the prevailing conditions of contract and that they are presented in a recognisable format that will facilitate their use.

Presented in a recognisable format

JCT98 with quantities requires bills of quantities to be prepared in accordance with the current version of the Standard Method of Measurement of Building Works 7th Edition, published by the Royal Institution of Chartered Surveyors and the Construction Confederation (SMM7R – at the time of publication).

The wider role

Even with the more or less universally adopted use of computers, the cost of producing bills of quantities is high and consequently their use as tender procurement and contract documents is often questioned and new techniques and procedures aimed at supplanting them are tried frequently. However, bills of quantities have survived thus far, no doubt because in addition to their two primary functions they contain vast amounts of information that can be of use in many ways. Also, they are frequently used in the preparation of interim valuations (see Chapter 14).

In preparing bills of quantities the quantity surveyor sifts through and processes much detailed information that, although not required by SMM7R to be included in the bills of quantities, may be of use to tendering contractors and the building team in connection with project planning, administration and the compilation of historical cost data in the form of elemental cost analyses. It is therefore worthwhile, in the early stages of the plan of work, to consider what additional information can be readily gathered in the measurement process and/or presented in the bills of quantities to assist tenderers and the building team to operate more effectively and efficiently.

Basic information

Bills of quantities are prepared by describing and quantifying, in accordance with the rules of SMM7R, the work shown on the drawings and given in the specification. Amongst other things SMM7R requires that *'Bills of Quantities shall fully describe and accurately represent the quantity and quality of the works to be carried out'*.

It is normal practice to arrange bills of quantities in three generic sections that between them *'define the precise nature and extent of the required work'* (SMM7R General Rule 1.1). These are:

- preliminaries
- preambles
- measured works.

The **preliminaries** are normally arranged in the Uniclass preliminaries/

general conditions sections (see Chapter 6). The aim is to provide a comprehensive schedule of items relating to the project generally, the contract, the employer's requirements, the contractor's general cost items and work by others that can be individually priced by the contractor for the purposes of calculating their lump sum price and for fixing rates that can be subsequently used in the valuation of variations. Within this framework it is necessary to provide at least the following information:

- name, nature and location of the project
- names and addresses of the employer and consultants
- list of drawings and other tender and contract documents
- site boundaries, existing buildings and existing mains services
- description of the works, including the dimensions and shape of each building
- form of contract, including a schedule of standard clause headings, details of any amendments to standard conditions and appendix insertions
- employer's requirements or limitations relating to tendering, sub-letting, supply, provision, content and use of documents, management of the works, quality standards/control, security, safety, protection, method of working, sequence of work, timing, use of site, temporary facilities, temporary works, temporary services and operation/maintenance of the finished building
- contractor's general cost items relating to management, staff, site accommodation, services, facilities, mechanical plant and temporary works
- work or products by others directly employed by the employer
- prime cost sums for work by nominated sub-contractors (see Chapter 8) together with items for main contractor's profit and attendances
- prime cost sums for goods to be supplied by nominated suppliers (see Chapter 8) together with items for main contractor's profit
- provisional sums for work by local authorities and statutory undertakers
- provisional sums for the labour, materials and plant elements of dayworks
- provisional sums for defined and undefined provisional work.

It is important to recognise the difference between prime cost (pc) sums and provisional sums. Prime cost sums are not defined in either JCT98 or SMM7R. They are, however, used to make cost provision within the bills of quantities for works to be carried out by nominated sub-contractors or for goods to be supplied by nominated suppliers. Frequently the firms to be nominated will have been selected and their prices fixed by the design team at pre-tender stage, hence the pc sums included in the bills of quantities should be fairly accurate.

Provisional sums are defined in both JCT98 and SMM7R. They are used to make cost provision within the bills of quantities for works to be carried out by local authorities or statutory undertakers, dayworks and other defined and undefined provisional work that cannot be described and given in items in accordance with the rules of SMM7R. Also JCT98 requires the architect to issue instructions in regard to the expenditure of provisional sums.

General rule 10 of SMM7R clearly sets down what is meant by defined and undefined provisional work and this rule is repeated in JCT98 with quantities. Provided all of the required information is given in the description of defined provisional work attached to a provisional sum, the contractor is deemed to have made due allowance for that work in programming, planning and pricing preliminaries. However, where all of the required information is not given, the work is deemed to be undefined provisional work and the contractor will be deemed not to have made any allowance for that work in programming, planning and pricing preliminaries. It is therefore of utmost importance that provisional sums for undefined provisional work include due allowance to cover the costs deemed not to have been included by the contractor in his lump sum price. It is also worthy of note that when the nomination of a supplier or sub-contractor or work by local authorities or statutory undertakers occurs as a result of expenditure of a provisional sum, that sum will have to be sufficient to also cover the cost of the main contractor's profit, attendances and fixing only of any supplied items.

The **preambles** traditionally comprise clauses specifying the quality or standard of materials and workmanship required. However, the preambles now more commonly comprise a simple reference to a separate, comprehensive specification document (see Chapter 6) which is thereby incorporated into the bills of quantities and given contract document status. Either way these specification requirements are intended to reduce the need for long descriptions and repetition in the measured works section of the bills of quantities. It is not normal for items in the preambles to be directly priced. The cost of complying with them is usually included within the rates set against the individual items included in the measured works section.

The **measured works** section comprises a schedule of quantified descriptions of the constituent parts of a building project, including services installations and external works, compiled in accordance with SMM7R. Provided that the descriptions comply with the appropriate rules, the quantities may be approximated if they cannot be accurately determined at bill production stage. The items are normally arranged in the Uniclass work sections. The aim is to provide a comprehensive schedule of items, described and quantified according to industry-recognised conventions, which can be individually priced by the contractor for the purposes of calculating their lump sum price and for fixing rates that can be used subsequently in the valuation of variations.

Provided that the Uniclass classification referencing system is used to annotate the drawings, specifications and bills of quantities, items in the bills can be directly related through that referencing system to relevant items in the specification and details on the drawings.

Formats

For tender purposes it is normal to arrange bills of quantities in the Uniclass work sections, as Example 7/1. However, it may be desirable, for various reasons, to add further information and/or to rearrange the format. Provided that the quantity surveyor appropriately annotates all items when measuring the work, the resultant bills of quantities can be readily reformatted to provide, for example:

- **Locational bills of quantities**, as Example 7/2, in which the Uniclass work sections are still adopted but the quantity for each item is broken down and allocated to a particular position within the project (e.g. a particular building, a part of a building, a house type). This format was developed to assist in the more accurate pricing of items.
- **Annotated bills of quantities**, as Example 7/3, in which the Uniclass work sections are still adopted but each item is annotated as to what it is and where it is located within the project. The annotations can be provided in a separate document, in a separate section of the bill or, most usefully, facing each item on the back of the preceding page in the bill. This format was developed as an extension of the locational bill.
- **Elemental bills of quantities**, as Example 7/4, in which the sections are based on the functional elements of the building, normally those used by the Building Cost Information Service (BCIS) (see Chapter 4), with the work in each element being arranged in Uniclass work sections. This format was developed to facilitate elemental cost analyses and as an aid to cost planning.

The early selection and appointment of the management contractor contemplated by the JCT98 Management Contract enables the employer to benefit from the management expertise of the contractor at an early stage in the project. When using this type of contract the project is split into discrete 'packages' that are let by the management contractor as separate 'works contracts'. The precise number of packages and their demarcation are determined by the design team and the management contractor on a project-by-project basis. The management contractor's expertise and advice is relied upon to attain the most efficient way of splitting the project into packages. Care is needed in co-ordinating the interfacing of packages to ensure that items of work are neither overlooked nor duplicated. Complete bills of quantities may be prepared for each package using any or all of the formats described above.

The examples of different formats for bills of quantities as given below are not exhaustive. It is up to the design team in general and the quantity surveyor in particular to decide how best to format the bills of quantities for each project to maximise the benefit to all concerned. It is worth remembering that efficient management of documentation is a valuable aid to efficient working on site and that can be of cost benefit to the employer.

Note: Extracts from two separate sections covering three trades are set out below to allow comparison with other formats. The sections and items within them are in the same order as SMM7R.

						£
	F MASONRY **F10 Brick/Block walling**					
	Concrete blocks, BS6073, 440mm x 215mm, solid, keyed one side, in cement-lime mortar (1:1:6), stretcher bond					
	Walls					
A	100mm thick : vertical	427	m^2			
B	100mm thick : vertical : curved on plan to 1500mm radius	12	m^2			
C	200mm thick : vertical	219	m^2			
	M SURFACE FINISHES **M20 Plastered/Rendered/Roughcast coatings**					
	Plaster : BS1191 part 2, undercoat 11mm thick, finishing coat 2mm thick : steel trowelled					
	Walls					
H	width >300mm : 13mm thick two coat work on brickwork or blockwork	860	m^2			
	Ceilings					
J	width >300mm : 13mm thick two coat work on concrete	435	m^2			
	Isolated columns					
K	width ≤300mm : 13mm thick two coat work on concrete	22	m			
	M60 Painting/Clear finishing					
	One coat sealer : two coats emulsion paint, matt finish					
	General surfaces					
T	girth >300mm	1295	m^2			
U	isolated surfaces girth ≤300mm	22	m			

Example 7/1   Uniclass work section bills of quantities.

Note: Extracts from two separate sections covering three trades are set out below to allow comparison with other formats. The sections and items within them are in the same order as SMM7R. The quantity of each item is broken down and allocated to a particular position within the project here represented by the letters A, B & C.

						£
	F MASONRY					
	F10 Brick/Block walling					
	Concrete blocks, BS6073, 440mm x 215mm, solid, keyed one side, in cement-lime mortar (1:1:6), stretcher bond					
	Walls					
A	100mm thick : vertical					
	A 340 + B 67 + C 20	427	m^2			
B	100mm thick : vertical : curved on plan to 1500mm radius					
	A 0 + B 3 + C 9	12	m^2			
C	200mm thick : vertical					
	A 99 + B 57 + C 63	219	m^2			
	M SURFACE FINISHES					
	M20 Plastered/Rendered/Roughcast coatings					
	Plaster : BS1191 part 2, undercoat 11mm thick, finishing coat 2mm thick : steel trowelled					
	Walls					
H	width >300mm : 13mm thick two coat work on brickwork or blockwork					
	A 688 + B 129 + C 43	860	m^2			
	Ceilings					
J	width >300mm : 13mm thick two coat work on concrete					
	A 350 + B 65 + C 20	435	m^2			
	Isolated columns					
K	width ≤300mm : 13mm thick two coat work on concrete					
	A 15 + B 0 + C 7	22	m			
	M60 Painting/Clear finishing					
	One coat sealer : two coats emulsion paint, matt finish					
	General surfaces					
T	girth >300mm					
	A 1038 + B 194 + C 63	1295	m^2			
U	isolated surfaces girth ≤300mm					
	A 15 + B 0 + C 7	22	m			

Example 7/2   Locational bills of quantities.

Note: Extracts from two separate sections covering three trades are set out out on the facing page to allow comparison with other formats. The sections and items within them are in the same order as SMM7R. Annotations are given below.

F   MASONRY
F10 Brick/Block walling

Annotations

A   Non-loadbearing partitions, first floor (drawing 456/78)

B   Stores, ground and first floor and staircase enclosure walls

C   Non-loadbearing partitions, ground floor (drawing 456/77)

M   SURFACE FINISHES
M20 Plastered/Rendered/Roughcast coatings

Annotations

H   Loadbearing partitions, ground floor

J   Soffit of first floor

K   Columns, entrance hall

M60 Painting/Clear finishing

Annotations

T   Loadbearing partitions ground floor and soffit of first floor (refer to colour schedule)

U   Columns, entrance hall (refer to colour schedule)

Example 7/3   Annotated bills of quantities. (*Continued.*)

Note: Extracts from two separate sections covering three trades are set out below to allow comparison with other formats. The sections and items within them are in the same order as SMM7R. Annotations are set out on the facing page.

			£
	F MASONRY **F10 Brick/Block walling**		
	Concrete blocks, BS6073, 440mm x 215mm, solid, keyed one side, in cement-lime mortar (1:1:6), stretcher bond		
	Walls		
A	100mm thick : vertical	427 m²	
B	100mm thick : vertical : curved on plan to 1500mm radius	12 m²	
C	200mm thick : vertical	219 m²	
	M SURFACE FINISHES **M20 Plastered/Rendered/Roughcast coatings**		
	Plaster : BS1191 part 2, undercoat 11mm thick, finishing coat 2mm thick : steel trowelled		
	Walls		
H	width >300mm : 13mm thick two coat work on brickwork or blockwork	860 m²	
	Ceilings		
J	width >300mm : 13mm thick two coat work on concrete	435 m²	
	Isolated columns		
K	width ≤300mm : 13mm thick two coat work on concrete	22 m	
	M60 Painting/Clear finishing		
	One coat sealer : two coats emulsion paint, matt finish		
	General surfaces		
T	girth >300mm	1295 m²	
U	isolated surfaces girth ≤300mm	22 m	

Example 7/3   *(Continued.)*

Note: Extracts from two separate elements covering three trades are set out below to allow comparison with other formats. The elements are set out in the same order as is used by BCIS; sections and items within each element are in the same order as SMM7R.

					£
	2.G INTERNAL WALLS AND PARTITIONS **F10 Brick/Block walling**				
	Concrete blocks, BS6073, 440mm x 215mm, solid, keyed one side, in cement-lime mortar (1:1:6), stretcher bond				
	Walls				
A	100mm thick : vertical	427	m²		
B	100mm thick : vertical : curved on plan to 1500mm radius	12	m²		
C	200mm thick : vertical	219	m²		
	3.A WALL FINISHES **M20 Plastered/Rendered/Roughcast coatings**				
	Plaster : BS1191 part 2, undercoat 11mm thick, finishing coat 2mm thick : steel trowelled				
	Walls				
H	width >300mm : 13mm thick two coat work on brickwork or blockwork	860	m²		
	M60 Painting/Clear finishing				
J	One coat sealer : two coats emulsion paint, matt finish				
	General surfaces				
K	girth >300mm	860	m²		
L	isolated surfaces girth ≤300mm	22	m		
	3.C CEILING FINISHES **M20 Plastered/Rendered/Roughcast coatings**				
	Plaster : BS1191 part 2, undercoat 11mm thick, finishing coat 2mm thick : steel trowelled				
	Ceilings				
R	width >300mm : 13mm thick two coat work on concrete	435	m²		
	M60 Painting/Clear finishing				
	One coat sealer : two coats emulsion paint, matt finish				
	General surfaces				
S	girth >300mm	435	m²		

Example 7/4   Elemental bills of quantities.

Chapter 8
Nominated Sub-Contractors and Nominated Suppliers

Sub-contractors

JCT98 is drafted on the premise that the contractor will be wholly responsible for carrying out and completing the works. However, as sub-contracting is customary in the construction industry, JCT98 makes provision for such arrangements by recognising two basic categories of sub-contractors to whom a part of the works may be sub-let:

- domestic sub-contractors
- nominated sub-contractors.

Domestic sub-contractors fall into two categories:

(1)   Provided that the contractor obtains the written consent of the architect, which consent must not be unreasonably delayed or withheld, he may sub-let work to any person of his choice and such a person is referred to as a domestic sub-contractor.

(2)   The employer may require certain items of work to be executed by one of not less than three persons named in the contract documents. In such a case the contractor must sub-let that work to one of the named persons and such person is also referred to as a domestic sub-contractor.

It is important to note that the contractor remains wholly responsible for carrying out and completing the work whether or not it is sub-let to a domestic sub-contractor.

Nominated sub-contractors are defined in JCT98 as sub-contractors to the contractor for the supply and fixing of particular materials or goods or for the execution of particular work, whose final selection and approval is reserved to the architect by agreement between the contractor and the architect on behalf of the employer or by use of a prime cost sum or by naming a sub-contractor in:

- the contract documents or
- any instruction on the expenditure of a provisional sum or
- any instruction requiring a variation for work additional to that shown on or described in the contract documents which is of a similar kind to that for which the contract documents provided that the architect would nominate a sub-contractor.

It is also a requirement that such sub-contractors are nominated in accordance with the relevant provisions in JCT98 and that no person against whom the contractor makes reasonable objection shall be nominated. The JCT98 procedure for nomination of a sub-contractor is discussed later in this chapter.

Nominated sub-contractors

As buildings become more complex and the systems within them more sophisticated, early decisions have to be made on various elements of them before the design team are able to proceed with the detailed design. For example, much thought must be given to the detail of a structural frame before the cladding of it can be designed; mechanical and electrical systems are frequently complex and make considerable demands on space for both plant and distribution that can affect storey heights and floor layouts. Thus it is essential for decisions relating to a number of functional elements of the building, including the selection of the principal specialist sub-contractors, to be made very early in the design process. Sub-contractors selected in this way are invariably required to contribute to or be responsible for some of the detailed design. This frequently happens in the case of structural steelwork, external wall cladding, mechanical and electrical systems, lifts and escalators.

This early involvement of sub-contractors in the design process creates a special relationship between them and the employer and in due course leads to the situation in which the employer requires the contractor to enter into a contract with the selected sub-contractor. The contractual arrangements for this sub-contract require careful consideration.

Early JCT standard forms of contract made special provisions for sub-contractors selected (nominated) in this way. Unfortunately those provisions had many shortcomings, both in terms of the contractual relationship of the parties and in the responsibility for any design input, giving rise to disputes which led to arbitration and litigation. Over the years, the nomination procedure, which places specific duties and responsibilities on the employer, the contractor, the sub-contractor and some other members of the design team, has been much refined such that the procedure for nomination of a sub-contractor prescribed in JCT98 is designed to resolve most issues that could otherwise lead to delays and disputes before the contractor and the nominated sub-contractor are instructed to enter into a contract.

*... the contractor and the nominated sub-contractor are instructed to
enter into a contract*

JCT98 sets out in detail the procedure to be followed and lists the standard
documents issued by JCT to be used when nominating sub-contractors. The
documents listed are:

- **NSC/T** – The Standard Form of Nominated Sub-Contract Tender 1998 Edi-
  tion comprising:
  Part 1 – The Employer's Invitation to Tender to a Sub-Contractor
  Part 2 – Tender by a Sub-Contractor
  Part 3 – Particular Conditions (to be agreed by the Contractor and the Nomi-
  nated Sub-Contractor)
- Agreement NSC/A – The Standard Form of Articles of Nominated Sub-
  Contract Agreement between a Contractor and a Nominated Sub-Contrac-
  tor, 1998 Edition
- Conditions NSC/C – The Standard Conditions of Nominated Sub-Contract,
  1998 Edition, incorporated by reference into Agreement NSC/A
- Agreement NSC/W – The Standard Form of Employer/Nominated Sub-
  Contractor Agreement, 1998 Edition
- Nomination NSC/N – The Standard Form of Nomination Instruction for a
  Sub-Contractor.

The procedure for using these documents comprises the following four stages:

(1)　The architect sends his enquiry to tendering sub-contractors together with:
　　(a) NSC/T Part 1 completed by the architect with relevant main contract and sub-contract information
　　(b) NSC/T Part 2 for completion by the tendering sub-contractors
　　(c) NSC/W with page 1 completed apart from the date, for signing/ entering into as a deed by the tendering sub-contractors
　　(d) the numbered tender documents (e.g. drawings, specification, bills of quantities, health and safety plans).

(2)　The tendering sub-contractors complete NSC/T Part 2, part complete NSC/W by signing/entering into as a deed and return both documents to the architect together with any further documents considered necessary to support their tender.

(3)　The architect selects the appropriate sub-contractor and has the employer approve and countersign their NSC/T Part 2 and execute their partially completed NSC/W. The architect then completes NSC/N and issues it to the contractor, with a copy to the sub-contractor, together with:
　　(a) NSC/T Part 1 completed by the architect
　　(b) NSC/T Part 2 completed by the sub-contractor and approved and countersigned by the employer
　　(c) the numbered tender documents and any additional documents approved by the architect
　　(d) NSC/W completed by the sub-contractor and the employer
　　(e) confirmation of any alterations to the information given in NSC/T Part 1 relating to obligations or restrictions imposed by the employer, order of works, employer's requirements and type and location of access
　　(f) principal contractor's health and safety plan.
　　NSC/T Part 2 provides that the nominated sub-contractors may withdraw their tender when they receive written notification of the identity of the main contractor.

(4)　Within ten days from the receipt of the architect's instruction the contractor:
　　(a) completes NSC/T Part 3 in agreement with the sub-contractor
　　(b) executes Agreement NSC/A with the sub-contractor
　　(c) sends copies of the executed Agreement NSC/A and the agreed and signed NSC/T Part 3 to the architect.

Since it is most unlikely that the contractor and the sub-contractor will have discussed the matter prior to stage 4, it may well be that they are unable to conclude an agreement in the ten days prescribed. When this situation arises the contractor is required to give written notice to the architect informing him either of the date by which he expects to have satisfactorily concluded stage 4 or of those matters that are preventing him from concluding stage 4. Within a reasonable period of time of receiving such a notice the architect must act upon it. In the former case he must consult with the contractor and, if he considers it reasonable, fix a later date for the conclusion of stage 4. In the latter case he must inform the contractor, in writing, either that the matters notified do not justify non-compliance, in which case the contractor shall complete stage 4, or if the matters notified do justify non-compliance he has to:

- issue further instructions that will facilitate compliance or
- cancel the nomination instruction and either omit the work or nominate another sub-contractor.

It is worthy of note here that any instruction issued by the architect under the JCT98 conditions relating to nominated sub-contractors may be grounds for an extension of time, provided that it is considered it will cause a critical delay. However, such instructions do not give rise to an entitlement to reimbursement of loss and/or expense.

Obtaining tenders

Although the provisions of JCT98 in respect of nominated sub-contractors are rather complicated, they do go a long way towards eliminating many of the problems that have arisen in the past in connection with the contractual arrangements for nomination.

Whilst the use of standard forms simplifies the process of obtaining tenders and ensures that all firms tendering do so on the same contractual terms, these forms do not, by themselves, ensure the submission of good tenders. For this to be achieved the standard forms need to be accompanied by sufficient additional information to fully apprise the tenderers of what is required of them. This additional information will comprise some, if not all, of the following documents:

- drawings
- specification
- bills of quantities/schedules of rates
- health and safety plans.

Specifications and bills of quantities for sub-contract works must not contain prime cost sums as the contract neither provides for their omission nor empowers the architect to issue instructions in regard to their expenditure.

Design by the nominated sub-contractor

JCT98 does not contemplate design by the contractor. Where there is a requirement for the contractor to design part of the works JCT has produced a Contractor's Designed Portion Supplement for use with JCT98. Where the contractor is required to design the whole of the works, the appropriate form of JCT contract to use is JCT98 With Contractor's Design. However, neither of these documents is within the scope of this book since it is based on the traditional system whereby the employer commissions a design team to design his required building and employs a contractor to build to that design.

Whether or not the main contract contemplates design by the main contractor, design input is frequently required from nominated sub-contractors. Where such design input is properly planned, co-ordinated with the other production information and made available to the main contractor on time, as good practice dictates, there are seldom any practical problems. However, when this does not happen and disruption and/or delays result, the employer is likely to suffer loss of time and money without having any remedy under JCT98. In recognition of this situation JCT has, via Agreement NSC/W, created a direct contractual link, and thereby a route for redress, between the employer and the nominated sub-contractor.

JCT98 and Agreement NSC/W

When using JCT98, Agreement NSC/W is an integral part of the nomination procedure. By way of this agreement the nominated sub-contractor warrants to exercise all reasonable skill and care in his:

- design of the nominated sub-contract works
- selection of materials and goods for the nominated sub-contract works
- satisfaction of any performance specification or requirements.

NSC/W provides for this warranty to survive even when no sub-contract is entered into by the main contractor and the nominated sub-contractor. The nominated sub-contractor also undertakes:

- to supply all information, drawings and details timeously so as not to give cause for a valid claim for an extension of time and/or reimbursement of direct loss and/or expense by the main contractor
- not to default on his obligations under the nominated sub-contract so as to avoid the issue of an instruction to determine his employment
- to perform the nominated sub-contract in such a way that the main contractor will not become entitled to an extension of time as a result of a delay by the nominated sub-contractor

- to indemnify the employer against any direct loss and/or expense arising from a renomination necessitated by the default of the nominated sub-contractor.

By way of the same agreement the employer undertakes:

- to pay the nominated sub-contractor for the design, purchase or fabrication of components ordered before the issue of the instruction on Nomination NSC/N
- that the architect will direct the main contractor and inform the nominated sub-contractor of the amount of any interim or final payment included in interim certificates in respect of the nominated sub-contract works
- that the architect will operate the provisions of JCT98 relating to the final payment of nominated sub-contractors
- that the architect will operate the provisions of JCT98 relating to the direct payment of nominated sub-contractors by the employer.

Agreement NSC/W also contains provisions relating to adjudication, arbitration and legal proceedings. These matters are considered in Chapter 16.

Nominated suppliers

Nominated suppliers are defined in JCT98 as suppliers to the contractor of materials or goods, for fixing by the contractor, who have been nominated by the architect in one of the following ways:

- A prime cost sum for the supply of materials or goods is included in the contract documents and the supplier of those materials or goods is either named in the contract documents or is named by the architect in an instruction issued for the purpose of nominating a supplier.
- A provisional sum is included in the contract documents and the instruction in regard to the expenditure of that sum makes the supply of materials or goods the subject of a prime cost sum and the supplier of those materials or goods is either named in that instruction or is named by the architect in another instruction issued for the purpose of nominating a supplier.
- A provisional sum is included in the contract documents and the instruction in regard to the expenditure of that sum makes the supply of materials or goods, for which there is only one source of supply, the subject of a prime cost sum (the sole supplier is deemed to have been nominated by the architect).
- A variation required by the architect specifies materials or goods for which there is only one source of supply and the supply of such materials or goods

is made the subject of a prime cost sum (the sole supplier is deemed to have been nominated by the architect).

However, when materials or goods specified in the contract documents are available from only one supplier or when the supplier of materials or goods specified in the contract documents is named but a prime cost sum for the supply of those materials or goods is not included in the contract documents, neither the sole supplier nor the named supplier are nominated suppliers. If the nomination of a supplier arises as the result of the expenditure of a provisional sum or of an instruction requiring a variation, the supply of the materials or goods concerned has to be made the subject of a prime cost sum.

The design team would be well advised to consider whether or not nominated suppliers will be required as early in the pre-contract stage as possible so that appropriate provision can be made in the contract documents.

Chapter 9
Obtaining Tenders

Introduction

It is frequently desirable to appoint the contractor at an early stage in the design process but to do so necessitates various modifications to traditional tendering procedures. Such matters are considered in the Aqua Group's book *Tenders and Contracts for Building.* Having said that, traditional tendering procedures are still widely adopted and are therefore the subject of this chapter.

There have so far been two significant published codes relating to the selection of contractors. In January 1996 the National Joint Consultative Committee for Building (NJCC) published *Code of Procedure for Single Stage Selective Tendering* in collaboration with the Scottish Joint Consultative Committee and The Joint Consultative Committee for Building, Northern Ireland. The NJCC was voluntarily dissolved on 27 June 1996 but there is still a tendency within the construction industry to use its code. In May 1997 the Construction Industry Board (CIB) published *Code of Practice for the Selection of Main Contractors.* The CIB ceased to exist on 29 June 2001 but its code, which does not seem to be as widely adopted as the NJCC code, continues to be published by Thomas Telford Publishing.

Both the NJCC and the CIB publications codify the principles of good practice relating to the selection of contractors and although they differ in detail they both cover the following topics:

- tender list
- tendering procedure:
  - preliminary enquiry
  - tender documents and invitation
  - tender period
  - tender compliance
  - late tenders.
- tender assessment:
  - opening tenders

***The primary aim … is to produce a list of contractors all of whom are
capable of completing the job satisfactorily***

- examination and adjustment of the priced document
- negotiated reduction of tender.
- notification of results.

The primary aim of selection is to produce a list of contractors all of whom are
capable of completing the project satisfactorily. The final choice of contractor
can then be the one submitting the lowest bona fide tender.

Tender list

The cost to a contractor of preparing a tender is high and is therefore reflected
in the cost of building. The greater the number of contractors tendering for any
one project, the higher will be the abortive cost. It therefore follows that open
tendering, which invariably leads to long tender lists, unnecessarily increases
the cost of building and should be avoided.

In order to minimise abortive tendering costs the design team should ensure, as early in the design process as possible, that there exists a list of suitable
contractors from which those invited to tender will be selected during stage H
of the plan of work (see Chapter 2). Those employers who build frequently will
normally maintain a list of suitable contractors. Such a list should be reviewed
on a regular basis so that contractors who have not performed well or who
have otherwise become unsuitable can be removed from the list and those not
on the list who meet the relevant criteria can be added to it. When a list of approved contractors does not exist, one should be compiled from firms known
to the employer and the design team and/or from those who respond to press
advertisements. Such advertisements need to be carefully worded and must

at least indicate the size, nature and location of the project and the date when the tender documents will be ready.

It is worth noting here that European Union procurement rules set down contractor selection and contract award procedures applicable to public authorities in respect of contracts above certain specified values. Such contracts must be the subject of notices in the *Official Journal of the European Communities*. Provided that the contracting authority opts for what is termed the 'restricted procedure' then, by use of the criteria stated in the notice, it may select those who will be invited to tender from the contractors who respond. The final list of contractors to be invited to tender should comprise no more than six plus two reserves. The main criteria for their selection should include:

- adequacy of available resources
- adequacy of technical and management structure
- financial stability and insurance cover
- health and safety record
- quality of work and adequacy of quality control
- performance record.

Preliminary enquiry

About a month before the tender documents are due to be despatched, the selected contractors should be sent a letter containing as much of the information listed below as possible and asking whether they wish to tender for the project:

- project name, function and general description
- employer
- design team
- location of site
- approximate cost range
- number of tenderers
- form of contract and whether to be executed under hand or as a deed
- any nominated sub-contractors for major items
- anticipated date of possession
- contract period
- anticipated date for despatch of tender documents
- tender period
- period for which tender is to remain open.

After the latest date for response to the preliminary enquiry the list of tenderers should be finalised and those included on the list notified. At the same time any contractors who asked to tender by responding to an advertisement and who are not included on the tender list should be informed.

Tender documents and invitation

On the date named in the preliminary enquiry the tender documents should be sent to or made available for collection by the selected tenderers. The tender documents should comprise:

- a checklist of all tender documents
- instructions to tenderers
- two copies of the drawings, schedules and specification
- two copies of the bills of quantities or pricing schedules
- two copies of the health and safety plan
- two copies of the form of tender
- a suitably addressed envelope, for the return of the tender, endorsed with the word 'Tender' and the name of the project together with the latest date and time for the tender return
- a suitably addressed envelope, for the return of the priced bills or schedules in support of the tender (if they are to be returned with the tender), endorsed with the project name and the words 'Priced Bills' or 'Priced Schedules' together with a space for the tenderer's name.

The instructions to tenderers referred to above should include:

- where and by when to submit the tender
- any information required to be submitted with the tender, such as method statement, quality control resources and programme of work
- method of packaging and identifying the tender
- method for dealing with any queries relating to the tender documents
- method of dealing with any errors or inconsistencies in the tender documents discovered after they have been issued
- whether or not alternative proposals are acceptable if accompanying a compliant tender
- notes relating to any performance-specified work and/or contractor's designed portion
- required period of validity for the tender
- anticipated tender acceptance date
- arrangements for inspecting additional information
- arrangements for visiting the site
- whether the employer will accept the lowest or any tender
- tender assessment criteria
- method of handling errors in tenders
- method of communication of tender results.

The contractor should of course acknowledge receipt of the documents and confirm that he is still able to provide a bona fide tender by the due date or, if for some reason he is unable to do so, he should return the documents immediately.

Tender period

The time required by a contractor to prepare a tender is dependent on both the size and complexity of the project. Whilst it is generally accepted that the minimum tender period should be four weeks, a longer period will be required in some instances. If realistic prices are to be tendered it is imperative that tenderers be given sufficient time in which to obtain competitive prices from their suppliers and sub-contractors, to comply with the requirements of CDM94 and to properly formulate their bids.

Tender compliance

In order to achieve fair competitive tendering it is essential that any unauthorised amendments to or qualifications of the tender documents by a tenderer render his tender non-compliant and subject to rejection. However, the tenderer should be given the opportunity to withdraw the amendments/qualifications and stand by his tender. It is also essential that unsolicited alternative bids, either in terms of price or time, are considered non-compliant and rejected.

Late tenders

The latest time and date for the submission of tenders is best established when the tender documents are first made available. This date may, for a variety of reasons, have to be extended, in which case all tenderers must be notified. Any tender received after the latest time and date fixed for their receipt should not be admitted to the competition and is best returned, unopened, to the tenderer.

Opening tenders

Tenders should be opened as soon as possible after the published time for their receipt. It is as well if at least two people are present. Upon opening, each tender should be countersigned by two of those present and the tenderer and tendered amount should be entered on a list which, when complete, should also be signed by two of those present.

Examination and adjustment of the priced document

The lowest tenderer should be requested to provide a priced document (bills of quantities or pricing schedule) in support of his tender if this was not required to be submitted with the tender. The quantity surveyor should examine the priced document in support of the lowest tender to determine that any amendments notified during the tender period have been properly included and to detect any errors in computation. Any errors found are dealt with in the manner prescribed in the instructions to tenderers.

The two most frequently adopted methods for dealing with errors in computation are:

(1) The tenderer is advised of the errors and is given the opportunity to confirm or withdraw his offer. If he withdraws, the examination process is repeated with the next lowest tenderer. If he confirms, the pricing document is endorsed to the effect that all rates and prices, excluding those relating to preliminaries, provisional sums and prime cost sums, be considered increased or decreased in the same proportion as the corrected total of the priced items exceeds or falls short of the uncorrected total of such items. It is these adjusted rates and prices that will subsequently be used for valuing variations.

(2) The tenderer is advised of the errors and is given the opportunity to amend his tender to correct genuine errors or to withdraw his offer. If he amends and the amended tender is no longer the lowest, or if he withdraws, the examination process is repeated with the lowest tenderer.

... is prepared to stand by his tender

When a tender appears to be free from error, or when in spite of error the contractor is prepared to stand by his tender, or when a tender is still lowest after amendment, such a tender should be recommended for acceptance. However, there are occasions when the tender instructions indicate that price will not be the only criterion by which the offers are to be assessed and that such factors as the tenderer's proposed solutions to specific problems, his proposed method of working or his proposed programme will be equal to or of greater importance than price. When this is the case then, following their evaluation of the bids against the relevant criteria, the design team may conclude that the lowest price offered is not necessarily the most appropriate one to recommend for acceptance.

Negotiated reduction of a tender

Good practice dictates that a contractor's tendered price should not be changed on an unamended scope of works except for the correction of genuine errors, as noted above. However, should the lowest tender exceed the employer's budget, it is permissible for a reduced price to be negotiated with the lowest tenderer on the basis of agreed changes in the specification and/or quantity of the work. If such negotiations fail then, and only then, is it permissible to commence negotiations with the next lowest tenderer.

Notification of results

Immediately after the tenders have been opened all but the lowest three tenderers should be notified that their tenders have not been successful. At the same time the second and third lowest should be advised that they are not the lowest but that they may be approached again should the lowest tenderer withdraw his offer. As soon as a decision has been made to accept a tender, all unsuccessful tenderers should be so notified and any submitted priced documents should be returned to them unopened.

The employer should be encouraged to decide as quickly as possible which tender to accept, for it is of the utmost importance to a contractor to know whether or not his tender has been successful. Once the contract has been let it is good practice to provide each tenderer with a complete list of the tender prices received. Whilst this does not disclose who tendered which amount, it does enable each tenderer to see how he stood in relation to the rest.

Tender analysis

Frequently it will be required for the quantity surveyor to analyse the successful tender – and in any event it is good practice. Traditionally the analysis is based on the functional elements of the building (see Chapter 4) to enable the quantity surveyor to revise his cost plan to reflect the tendered prices in order to establish a proper basis for post-contract cost control (see Chapter 13) and for the purposes of adding to the industry's historical cost data bank.

More and more projects are dependent upon third-party funding and government-sponsored financial incentives for their economic viability. The quantity surveyor is therefore frequently required to provide financial information, normally gleaned from his tender analysis, to satisfy funders, to support grant applications and to enable the employer to take advantage of capital allowances and other forms of tax relief offered by the Inland Revenue and HM Customs and Excise (see Chapter 17).

Chapter 10
Placing the Contract

The placing of the contract is a relatively simple, routine matter but the events that immediately precede it and those that immediately follow it are far from simple and are certainly of great importance.

Preparing and signing the contract documents

If the date of possession of the site by the contractor and the date for completion of the works have not previously been decided they should he agreed with the contractor whilst the contract documents are being prepared for signing. At the same time it is as well to make arrangements for the initial site meeting and to ensure that the contractor has no valid objection to any of the proposed nominated sub-contractors and suppliers (see Chapter 8).

The contract documents, which should be so labelled, will normally comprise the articles of agreement, the conditions of contract, the drawings showing the work to be done, the priced bills of quantities or the priced document and any post-tender negotiation documentation. In preparing JCT98 for signing it is necessary to complete the articles of agreement at the beginning and the appendix together with its annex relating to the terms of bonds at the end. It is also necessary to make the following deletions and amendments in the text of the conditions of contract:

- Clause 1.3 – amend as necessary the definition of public holidays if different public holidays are applicable.
- Clause 1.10 – amend as necessary when the parties to the contract do not wish the law applicable to the contract to be English law.
- Clauses 5.3.1.2 and 5.3.2 – delete clause 5.3.1.2 if the contractor is not required to provide a master programme and also delete the words in parentheses that relate to the master programme in clause 5.3.2.
- Clauses 22A, 22B and 22C – always delete two of these clauses. Clause 22A is applicable to the erection of new buildings when the contractor is required

to take out the insurance. Clause 22B is applicable to the erection of new buildings when the employer is required to take out the insurance. Clause 22C is applicable to alterations and extensions to existing structures in which case the contract provides only that the employer is required to take out the insurance. Footnotes to these clauses remind that it is sometimes not possible to obtain insurance in the precise terms required by JCT98, in which case the conditions must be amended accordingly.

- Clause 35.13.5.3.4 – amend to refer to the events on the happening of which bankruptcy occurs when the contractor is a person subject to bankruptcy laws and not the law relating to the insolvency of a company.

The information necessary for the completion of the articles of agreement and the appendix together with details of the above noted deletions and amendments should have been included in the tender documents. It is imperative that no other insertions, deletions or amendments are made to JCT98 without the prior agreement of the contracting parties.

There are three supplements to JCT98 which, if relevant, will have to be included in the documentation. These are:

- Sectional Completion Supplement
- Contractor's Designed Portion Supplement
- Composite Contractor's Designed Portion and Sectional Completion Supplement.

It is worth noting here that sectional completion is not to be confused with partial possession by the employer. Sectional completion is provided for by a supplement to JCT98 and relates to a specific requirement of the employer, recorded in the contract documents, for the works to be completed in discreet, phased sections by pre-determined dates. Partial possession by the employer, however, is covered by JCT98 clause 18 which provides that the employer may, with the consent of the contractor, take possession of a part or parts of the works at any time prior to practical completion.

The **Sectional Completion Supplement** contains modifications to JCT98 for use when the works are to be completed in phased sections, of which the employer will take possession upon practical completion of each. If this supplement is to be used it is necessary for the works to be divided into the sections in the tender documents at tender stage. Briefly this supplement comprises:

- instructions on the incorporation of the supplement into JCT98
- a replacement first recital
- a table of changes to the articles of agreement and the conditions
- a replacement annex.

The **Contractor's Designed Portion Supplement** contains modifications to

Contractor's Designed Portion ...

JCT98 for use when the contractor is required to complete the design of a portion of the works. Briefly this supplement comprises:

- instructions on the incorporation of the supplement into JCT98
- replacement recitals
- a replacement article 1
- a schedule of modifications to the conditions
- a supplementary appendix.

The use of this supplement creates the following additional contract documents:

- **The Employer's Requirements** – this document shows and describes the requirements of the employer in respect of that portion of the works to be designed by the contractor.
- **The Contractor's Proposals** – this document shows and describes the contractor's proposals for the design of the relevant portion of the works to meet the employer's requirements.
- **The CDP Analysis** – this document is an analysis of the portion of the contract sum relating to the portion of the works to be designed by the contractor.

The **Composite Contractor's Designed Portion and Sectional Completion Supplement** is, as its name implies, simply both of the above described supplements brought together in one document for ease of use when both are to apply to any one contract.

Executing the contract

The tender documents should record, at tender stage, whether the agreement is to be executed under hand or as a deed. The Limitation Act 1980 provides that actions for breach of contract can be commenced within six years of the breach in respect of a contract executed under hand and within 12 years of the breach in respect of a contract executed as a deed. In the event of one party executing the contract under hand and the other party executing it as a deed, it would appear that the 12 years limitation period applies only against the party executing it as a deed.

The attestation clauses in JCT98 make provision for the agreement to be executed under hand or as a deed both by a company or other body corporate and by an individual. Irrespective of how the agreement is executed, both parties to the contract should initial each and every contract document. Also, they should both initial all amendments made to JCT98 and its supplements, when used.

Performance bonds and parent company guarantees

A performance bond is a three-party agreement between the contractor, a surety and the employer that provides for the surety to pay a sum of money to the employer in the event of default by the contractor. The value of the bond normally provided is equal to 10% of the original contract sum. As the raising of a bond invariably imposes a financial burden on the contractor in terms of borrowing, it is considered to be good practice for the bond to be released coincidentally with the achievement of practical completion.

The contractor will normally obtain such a bond from an insurance company or a bank. However, where the contractor is a subsidiary of a large organisation it may well be that a guarantee from the parent company will suffice instead of a bond. The bond holder and the terms of the bond or guarantee must be approved by the employer. Examples of a typical performance bond (see Example 10/1) and a typical parent company guarantee (see Example 10/2) are reproduced at the end of this chapter. Whether or not a bond or guarantee is required must be determined prior to tender stage and the appropriate entry made in the tender documents so that the tenderers can make due allowance for all associated costs in their tendered prices.

Collateral warranties

Collateral warranties in general and those to be provided by consultants in particular are considered in Chapter 2, where mention is made of the standard warranty agreements published by BPF.

Like consultants, main contractors are also frequently required to enter into collateral warranty agreements and are always required to do so when those entered into by the consultants are the BPF agreements CoWa/F and CoWa/P&T, under which the consultants' responsibilities are on the basis that the main contractor will provide contractual undertakings to the funding institution, purchaser or tenants to similar effect as those provided by the consultants. To provide for this, BPF published the following standard warranty agreements and enabling clauses in 1993:

- MCWa/F – Collateral Warranty for funding institutions by a main contractor and enabling clause 19B.
- MCWa/P&T – Collateral Warranty for purchasers and tenants by a main contractor and enabling clause 19A.

In October 2001 JCT published revised versions of these standard collateral warranty agreements and, at the same time, they published SCWa/F – Collateral Warranty for funding institutions by a sub-contractor and enabling clause (Y) for insertion in the sub-contract, and SCWa/P&T – Collateral Warranty for purchasers and tenants by a sub-contractor and enabling clause (X) for insertion in the sub-contract.

Issue of documents

JCT98 requires that the contract drawings and the contract bills or priced document remain in the custody of the architect or the quantity surveyor so as to be available for inspection by the contractor or the employer. It also requires that immediately after the contract has been executed the architect must provide the contractor with:

- one copy of the contract documents certified on behalf of the employer
- two further copies of the contract drawings
- two copies of the unpriced bills of quantities or specification/schedules of work.

JCT98 also requires that as soon as possible after the execution of the contract:

- the architect shall provide the contractor with two copies of any descriptive schedules or other similar documents necessary for use in carrying out the works
- when JCT98 clause 5.3.1.2 has not been deleted the contractor shall provide the architect with two copies of his master programme for the execution of the works.

Insofar as they are relevant to the project the following documents will also have to be provided to the contractor by the design team:

- party wall agreements
- condition surveys of adjoining properties
- conditional planning approvals
- tree preservation orders
- building control approval and notices to be served during the course of the works
- procedures and estimates for works to be carried out by statutory and local authorities.

The contractor is required by JCT98 to keep one copy of each of the following documents on site so as to be available to the architect:

- contract drawings
- the unpriced bills of quantities or specification/schedules of work
- descriptive schedules or other similar documents
- the master programme (unless clause 5.3.1.2 has been deleted)
- any further drawings or details issued to explain and amplify the contract drawings.

Insurances

The insurance provisions in JCT98 are to be found in clauses 21, 22, 22A, 22B, 22C and 22D.

Clause 21 requires the contractor to take out and maintain insurance against claims for personal injury to or death of any person or for loss, injury or damage to real or personal property that arises out of or in the course of or is caused by the carrying out of the works except for claims caused by a default of the employer or those for whom the employer is responsible. The insurance in respect of the contractor's liability to third parties must also extend to meet any like claims made against the employer by third parties. Insurance in respect of injury or death to an employee of the contractor must comply with the current legislation [at the time of writing this is the Employers' Liability (Compulsory

Insurance) Regulations 1989]. Insurance in respect of all other claims has to be for no less cover than is stated in the appendix to the contract.

Clause 21 also provides the employer with the option to require the contractor, by way of an architect's instruction, to insure (in the names of the employer and the contractor) against any expense, liability, loss, claim or proceedings incurred by the employer as a result of injury or damage to any property, other than the works and any site materials for the works, caused by:

- collapse
- subsidence
- heave
- vibration
- weakening or removal of support
- lowering of ground water

arising out of or in the course of or by reason of the carrying out of the works, excepting injury or damage:

- caused by the default of the contractor
- caused by errors or omissions in designing the works
- which can reasonably be foreseen to be inevitable
- which it is the responsibility of the employer to insure under JCT98 clause 22C (if applicable) – see below
- arising from the consequence of war and like acts
- caused by the excepted risks (as defined in JCT98 clause 1.3)
- caused by pollution or contamination
- resulting in damages payable by the employer for breach of contract.

The extent of cover required has to be stated in the appendix to the contract, but as this particular insurance is instigated by an architect's instruction the cost to the contractor of taking it out and maintaining it is added to the contract sum as an extra.

Clause 22 deals with insurance of the works and provides three alternatives:

- **Clause 22A** – where the works comprise new buildings and the contractor is required to take out and maintain a joint names policy for all risks insurance of the works.
- **Clause 22B** – where the works comprise new buildings and the employer is required to take out and maintain a joint names policy for all risks insurance of the works.
- **Clause 22C** – where the works are in or are extensions to existing structures the employer is required to take out and maintain a joint names policy for loss or damage caused to the existing structures and contents by one or

more of the specified perils and a joint names policy for all risks insurance of the works.

The term 'joint names policy' is expressly defined in JCT98 at clause 22.2 as being *'a policy which includes the Employer and the Contractor as the insured and under which the insurers have no right of recourse against any person named as an insured, or, pursuant to clause 22.3, recognised as an insured thereunder'*. JCT98 clause 22.3 further provides that the joint names policy must either provide for recognition of each nominated sub-contractor as an insured or include a waiver by the insurers of any right of subrogation which they may have against any such nominated sub-contractor. Subrogation is the doctrine under which an insurer, who has paid his insured for loss or damage suffered, is entitled to sue, in the insured's name, whoever caused the loss or damage.

The amount of cover to be provided by the joint names policy is the full reinstatement value of the works plus the percentage stated in the appendix to the contract to cover professional fees and, in the case of works in or extensions to existing structures, plus the full value of the existing structures and of their contents owned by or for which the employer is responsible.

JCT98 clause 22A.3 recognises that it is common practice for contractors to maintain annually renewable insurance policies which provide appropriate clause 22A all risks insurance and sets out deemed-to-satisfy provisions in respect of such annual policies.

JCT has recently issued Amendment 3 to JCT98 which, amongst other things, defines terrorism and terrorism cover and provides the employer with the option of determining the employment of the contractor or of requiring the contractor to complete the works at the employer's expense if insurers withdraw terrorism cover during the currency of the contract. These provisions were previously contained in JCT Amendment TC/94, issued April 1994. Following terrorist attacks on the UK mainland, particularly the IRA bombings in the City of London, which caused billions of pounds worth of damage, insurance against terrorism has become unavailable. In order to comply with the JCT98 requirement to maintain cover, the government agreed to act as 'reinsurer of last resort'. Information relating to terrorism cover and the position of the government as reinsurer of last resort is given in JCT Practice Note 3 (Series 2).

Clause 22D provides the employer with the option to require the contractor to insure against delays to completion of the works caused by any of the specified perils defined in clause 1.3 and which result in the architect granting an extension of time. The amount of cover will be at the rate for liquidated and ascertained damages and for the period given in the appendix to the contract. This type of insurance is expensive and will generally be in respect of delays of no more than ten weeks.

Construction insurance is an extremely complex subject which is not helped by the fact that available policies do not incorporate standard wording. Consequently the employer should always be advised to consult a specialist insurance adviser. Once insurances are in place it is important that they are maintained and renewed when necessary. It is therefore recommended that the design team checks that insurances are maintained and renewal premiums are paid when due. It is advisable to keep copies of premium receipts or brokers' confirmatory letters on file.

THIS BOND is made the .... day of .... 20 ......

BETWEEN ....

of ....

(hereinafter called 'the Contractor') of the first part.

and ....

of ....

(hereinafter called 'the Surety') of the second part.

and ....

of ....

(hereinafter called 'the Employer') of the third part.

1. By a Contract dated .... made between the Contractor and the Employer (hereinafter called 'the Contract') the Contractor has agreed to carry out the Works specified in the Contract for the sum of £ ....

2. The Contractor and the Surety are hereby jointly and severally bound to the Employer in the sum of £ .... (not exceeding ten per cent of the original Contract Sum) which sum shall be reduced by an amount equal to ten per cent of the value of any part or parts of the Works taken into possession of the Employer under the provisions of the Contract *(or of any Section of the Works upon the Architect/Contract Administrator certifying practical completion of that Section) provided that if the Contractor shall, subject to Clause 5 hereof, duly perform and observe all the terms, conditions, stipulations and provisions contained or referred to in the Contract which are to be performed or observed by the Contractor or if on default by the Contractor the Surety shall, subject to Clause 3 hereof, satisfy and discharge the damage sustained by the Employer thereby up to the amount of this Bond then this agreement shall be of no effect.

* *The wording in brackets should be deleted unless Sectional Completion applies.*

3. If the Contractor has failed to carry out the obligations referred to in Clause 2 hereof then written notice requiring the Contractor to remedy his failure, where possible, shall be given, and if the Contractor fails so to do or repeats his default or if the Contractor's employment under the Contract is determined in accordance with clause 27 of that Contract the Employer shall be entitled to call upon the Surety in accordance with Clause 2.

4. Any alteration to the terms of the Main Contract or any variations required under the Contract shall not in any way release the Surety from its obligations under this agreement.

5. The Contractor and the Surety shall be released from their respective liabilities under this agreement upon the date of the Practical Completion of the Works as certified by the Architect/Contract Administrator appointed under the Contract.

IN WITNESS whereof the parties hereto have executed this Document as a Deed the day and year first before written.

Contractor: ....

Signed by (insert name of Director) and   Director

(insert name of Company Secretary or   ....

second Director) for and on behalf of   Director/Company Secretary

....

Surety: ....

Signed by (insert name of Director) and   Director

(insert name of Company Secretary or   ....

second Director) for and on behalf of   Director/Company Secretary

....

Employer: ....

Signed by (insert name of Director) and   Director

(insert name of Company Secretary or   ....

second Director) for and on behalf of   Director/Company Secretary

....

Example 10/1   Performance bond.

THIS AGREEMENT is made the .... day of .... 20 ......

BETWEEN ....

whose registered office is at ....

(hereinafter called 'the Guarantor') of the one part and ....

whose registered office is at ....

(hereinafter called 'the Employer') of the other part,

WHEREAS

A.　　..

　　　　(hereinafter called ....

　　　　a subsidiary of the Guarantor has entered into a Contract with the Employer of even date to

　　　　carry out certain works at ....

　　　　(hereinafter called 'the Contract')

B.　　The Guarantor has agreed to guarantee the due performance by ....

　　　　........................ of the Contract in the manner hereinafter appearing:

NOW THE GUARANTOR HEREBY AGREES WITH THE EMPLOYER as follows:

1.　　If .... (unless relieved from the performance by any clause of the Contract or
　　　by statute or by the decision of a tribunal of competent jurisdiction) shall in any respect
　　　commit any breach of its obligations or duties under the Contract then the Guarantor
　　　will procure the remedying of any breach of .... obligations failing which
　　　the Guarantor will pay to the Employer and/or his assigns the amount (subject as
　　　hereinafter provided) of all losses, damages, costs and expenses which may be incurred
　　　by the Employer by reason of any default on the part of .... in performing its
　　　obligations and duties or otherwise to observe and comply with the provisions contained
　　　in the Contract to the extent that such losses, damages, costs and expenses are or would
　　　otherwise be recoverable by the Employer from .... under the said Contract.

2.　　The Guarantor further agrees with the Employer that it shall not in any way be released
　　　from liability hereunder by any alteration in the terms of the Contract made by
　　　agreement between the Employer and .... or in the extent or nature of the
　　　Works to be constructed, completed and maintained thereunder or by any allowance
　　　of time or forbearance or forgiveness in or in respect of any matter or thing concerning
　　　the Contractor or by any other matter or thing whereby (in the absence of this present
　　　provision) the Guarantor would or might be released from liability hereunder, and
　　　for the purposes of this Agreement any such alteration shall be deemed to have been
　　　made with the consent of the Guarantor and Clause 1 hereof shall apply in respect of
　　　the Contract so altered. Provided always that the amount of the Guarantor's liability
　　　under this Agreement shall be no greater than the amount which would have been
　　　recoverable against the Contractor by the Employer under the Contract and the same
　　　limitation periods fixed by statute which apply to the Contract shall equally apply to this
　　　Agreement.

IN WITNESS whereof the parties hereto have executed this Document as a Deed the day and year
first before written.

Guarantor:　　　　　　　　　　　　　　　　　　.....................................

Signed by (insert name of Director) and　　　　Director

(insert name of Company Secretary or　　　　　.....................................

second Director) for and on behalf of　　　　　Director/Company Secretary

....

Employer:　　　　　　　　　　　　　　　　　　.....................................

Signed by (insert name of Director) and　　　　Director

(insert name of Company Secretary or　　　　　.....................................

second Director) for and on behalf of　　　　　Director/Company Secretary

....

Example 10/2　Parent company guarantee.

Chapter 11
Meetings

Initial meeting

As soon as practicable after the contract has been placed, the building team should meet. Although this initial meeting may take place on site, it will more probably take place in the offices of the project manager or the architect or the main contractor.

The manner in which the initial meeting is conducted will greatly influence the success of the project, and succinct, clear direction from the chair will be a strong inducement to a similar response from others. Since at this stage the person having the most complete picture of the project is likely to be the project manager, if there is one, or the architect, if there is not, it is logical that the project manager or the architect should take the chair. The arrangements for this meeting should be discussed with the main contractor beforehand. It is suggested that as many of the following as are involved with the project should be requested to attend:

- employer
- project manager
- planning supervisor
- architect
- quantity surveyor
- structural engineer
- building services engineer(s)
- contractor
- principal nominated sub-contractors
- principal nominated suppliers
- clerk of works.

It is also suggested that the agenda for the initial meeting should include the following matters:

- introductions
- factors affecting the carrying out of the works
- programme
- nominated sub-contractors and suppliers
- lines of communication
- bonds, collateral warranties and insurances (see Chapter 10)
- financial matters
- procedure to be followed at subsequent meetings.

Introductions

The introduction of those attending needs no elaboration, although it is more than just a formality as it establishes the initial contact between individuals who must work together in harmony if the project is to run smoothly.

Factors affecting the carrying out of the works

These should be described fully in the contract documents but may require emphasis and clarification at the initial meeting. The factors may include:

- existing mains/services
- site investigation, including soils and ground water information
- access to the site
- parking restrictions
- use of the site
- risks to health and safety and health and safety plans
- work outside the site boundary
- concurrent work by others
- listed domestic sub-contractors
- equivalent products
- Considerate Constructors Scheme
- photographic records
- prohibited products
- protection of products
- samples
- building lines, setting out, critical dimensions and tolerances
- co-ordination of engineering services
- quality control
- use of explosives and pesticides and on-site burning of rubbish
- fire prevention
- protection of the works, existing retained features, adjoining buildings, work people and the public

- working hours and overtime working
- location of temporary accommodation, spoil heaps, etc.
- temporary fences, hoardings, screens, roads, name boards, etc.
- temporary services and facilities.

Programme

JCT98 clause 5.3.1.2 requires that the contractor shall provide the architect with two copies of his master programme for the execution of the works as soon as possible after the execution of the contract. Although a footnote in JCT98 advises that clause 5.3.1.2 should be deleted if no master programme is required, this should be done only in exceptional circumstances as experience has shown that a master programme is an essential management tool for both the contractor and the design team.

In addition to providing the master programme as soon as possible after the execution of the contract the contractor is also required to update it within 14 days of any decision by the architect that creates a new completion date. In any event the master programme should be reviewed regularly and updated whenever necessary. It is to be noted that the master programme is not a contract document and JCT98 clause 5.3.2 makes it clear that nothing contained in the master programme can impose any obligation on the contractor beyond the obligations imposed upon him by the contract documents. Some forms of contract do not require the contractor to provide a programme. In such circumstances the need for a programme should be stated in the bills of quantities or the specification.

In *'Glenlion Construction Limited -v- The Guinness Trust'* it was decided that:

(1)   The contractor is entitled to complete the works on or before the date for completion stated in the contract appendix, or any later completion date fixed pursuant to the appropriate conditions of contract; and

(2)   the contractor is not entitled to expect the design team to provide design information to enable him to complete the works before the date for completion stated in the contract appendix, or any later completion date fixed pursuant to the appropriate conditions of contract.

At first sight these two statements may appear to conflict with each other but with further thought it will be apparent that they do not. This is because, even if the contractor is capable of completing the works before the date for completion, it may not be possible for the design team to provide design information sooner than would be necessary for the contractor to complete the works by the contract completion date.

Prior to the initial meeting the contractor should determine, as far as possible, major material delivery dates from both domestic and nominated suppliers and required construction periods from both domestic and nominated sub-contractors. From this information and the Information Release Schedule, when provided by the employer (see JCT98 sixth recital and clauses 1.3 and 5.4), the contractor should formulate a draft master programme for the works which should then be circulated to all members of the building team. Whether or not an information release schedule has been provided, it is quite usual for the contractor to indicate on the draft master programme the latest dates by which drawings, schedules, instructions for placing orders and other information are required from the design team.

The draft master programme should be considered and, if necessary, adjustments agreed during the initial meeting. Following this the contractor should prepare his definitive master programme and circulate it to all concerned. An example of a simple bar chart programme is given as Example 11/1 at the end of this chapter.

The importance of adhering to dates once agreed, whether they be in respect of materials deliveries, execution of the work or the production of further necessary information by the design team, cannot be over-stressed.

Nominated sub-contractors and suppliers

Many contractual problems arise as a result of the architect failing to issue instructions in relation to the nomination of sub-contractors and suppliers timeously and as a result of poor communications between the main contractor and the nominated firms. It is therefore advisable at the initial meeting for the architect to inform the main contractor of his intended instructions relating to the nomination of sub-contractors and suppliers. It is also advisable to make it clear to the main contractor that, once nominated, the sub-contractors and suppliers are contractually his responsibility.

It is essential, for the smooth and efficient running of the project, that all nominations are made sufficiently early for the work concerned to be integrated into the contractor's programme of work without causing disruption. It is also essential that each member of the building team allow sufficient time in their work schedule to properly carry out the various complex and often time-consuming procedures involved in the selection and appointment of nominated sub-contractors and suppliers. Following the correct procedures and providing the appropriate documentation when nominating firms is of the utmost importance and is dealt with in Chapter 8.

Lines of communication

It is important at the initial meeting for the chairman to emphasise the need for all members of the building team to abide by the formal lines of communication provided for in JCT98 and to make reference to the golden rules of communication given in Chapter 1.

Financial matters

It is desirable at the initial meeting for the quantity surveyor to run through the various procedures set down in JCT98 relating to financial matters. These are to be found in:

- Clause 3   Contract Sum – additions or deductions – adjustment – Interim Certificates
- Clause 6   Statutory obligations, notices, fees and charges
- Clause 9   Royalties and patent rights
- Clause 13   Variations and provisional sums
- Clause 13A   Variation instruction – Contractor's quotation in compliance with instruction
- Clause 15   Value added tax – supplemental provisions
- Clause 18   Partial possession by the Employer
- Clause 19   Assignment and sub-contracts
- Clause 22   Insurance of the Works
- Clause 24   Damages for non-completion
- Clause 26   Loss and expense caused by matters materially affecting regular progress of the Works
- Clause 27   Determination by Employer
- Clause 28   Determination by Contractor
- Clause 28A   Determination by Employer or Contractor
- Clause 30   Certificates and payments
- Clause 31   Construction Industry Scheme (CIS)
- Clause 34   Antiquities
- Clause 35   Nominated Sub-Contractors
- Clause 36   Nominated Suppliers
- Clause 37   Fluctuations.

It is worth stressing that no adjustment in respect of varied work will be made to the contract sum unless the matter is covered by an architect's instruction issued in accordance with the terms of the contract. Also, it is worth reminding the contractor of any other financial matters required by the contract documents, such as the provision by him of a cash flow forecast showing the gross valuation of the works at the date of each interim certificate.

Procedure to be followed at subsequent meetings

Formal site meetings are not to be confused with the architect's periodic site visits and the numerous meetings between the main contractor and others that are necessary to the progress of the works. The frequency of formal site meetings will vary with the size and complexity of the project, the stage of the job and any difficulties being encountered. It would be unusual for such meetings to be held less frequently than once every four weeks.

The requirements of the employer as to the nature, normal frequency, procedures and location for formal site meetings should be given in the contract documents. These meetings will be chaired by the project manager, if there is one, or the architect, if there is not, and will normally be attended by the design team, the main contractor and such sub-contractors as are requested to be present. The object of the formal site meetings is to provide a forum for the presentation by the contractor of his progress report which should include, at least:

- a progress statement by reference to the master programme for the works
- details of any matters materially affecting the regular progress of the works
- any requirements for further drawings, details and instructions.

It also provides an opportunity for the contractor to raise any queries he may have with the design team as a whole and to formally report on health and safety issues.

The project manager/architect should notify the rest of the design team and the main contractor of the dates and times for site meetings and should request those whose presence is required to attend. It should be the responsibility of the main contractor to invite to the site meetings representatives of those sub-contractors and suppliers whom he or the design team wish to be present. Due consideration should be given to the value of people's time, with care being taken not to invite to meetings those whose presence is not really necessary.

The project manager/architect and the main contractor should agree the agenda for each site meeting prior to the meeting. A standard form of agenda is useful as a model and an *aide memoire*. A typical site meeting agenda is given as Example 11/2 at the end of this chapter. The minutes of site meetings will normally be taken and distributed by the chairman of the meeting. They must be impartial and concise and accurately record all decisions reached and actions required. Typical site meeting minutes are given as Example 11/3.

It is worthy of note that a badly run site meeting can do untold harm and be a serious waste of time for all attending whereas a well-managed meeting can be a great aid to the smooth running of the project. Further, there is little doubt that well-run meetings maintain the team's interest in the project and impart a sense of involvement and urgency which is difficult to achieve by any other means.

... a great aid to the smooth running of the project

Contractor's meetings

The main contractor will normally hold meetings with his sub-contractors and suppliers shortly before the formal site meetings. Such meetings help facilitate the accurate reporting of progress and ensure that all sub-contractors' and suppliers' information requirements are met. They also provide an opportunity to establish requirements for holes, chases, recesses, fixings and the like before work is put in hand and thereby avoid conflict with other work.

Employer's meetings

The employer may also hold separate meetings with the rest of the design team during the progress of the works. Such meetings are normally used to discuss general progress and to resolve any outstanding design issues, such as colour schemes. They also provide an opportunity to consider the possible effects on the works, and on the anticipated final costs, of proposed variations.

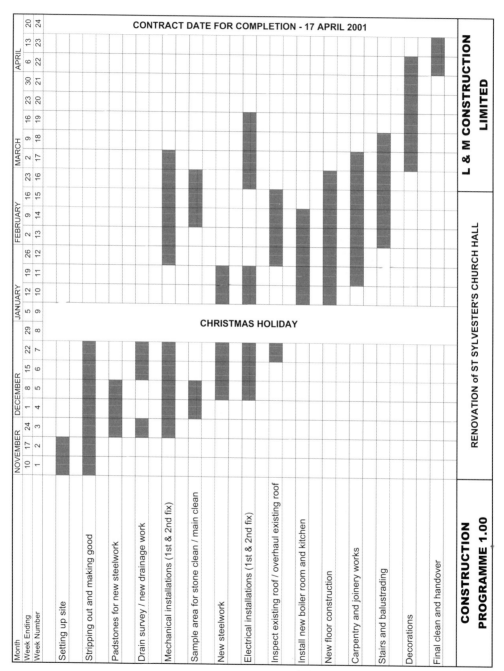

Example 11/1   Construction programme.

Project:	Shops & Offices, Newbridge Street, Borchester
Project ref:	456
AGENDA FOR SITE MEETING	
Date:	7 October 2002 at 10.00am

1.0	Apologies
2.0	Minutes of last meeting
3.0	Contractor's report General report Progress statement by reference to the master programme for the works Details of any matters materially affecting regular progress of the works Information received since last meeting Requirements for further drawings, details and instructions Health and safety matters
4.0	Clerk of Works' report Site matters Quality control Lost time
5.0	Design Consultants' reports Architect Structural Engineer Building Services Engineer
6.0	Quantity Surveyor's report
7.0	Contract completion date Assess likely delays Review factors from previous meeting List factors for review at next meeting Record anticipated completion date
8.0	Any other business
9.0	Date and time of future meetings Site meetings Site visits

Distribution:

Copies	2 Employer	3 Main Contractor	1 Project Manager
	1 Planning Supervisor	1 Architect	1 Quantity Surveyor
	1 Structural Engineer	1 Services Engineer	1 Clerk of Works

Example 11/2   Typical site meeting agenda.

Project:	Shops & Offices, Newbridge Street, Borchester	Ivor Barch
Project ref:	456	Associates Prospects Drive Fairbridge

SITE MEETING NO 4

Date:	7 October 2002
Location:	Site

Present:	S Gilbert	Employer (Aqua Products plc)
	A Morley	Main Contractor (L & M Construction Ltd)
	G Mackay	Main Contractor (L & M Construction Ltd)
	B Hunt	Project Manager
	Ivor Barch	Architect (Ivor Barch Associates)
	R W Pipe	Quantity Surveyor (Fussedon Knowles & Partners)
	I Tegan	Structural Engineer (GFP & Partners)
	F Adams	Services Engineer (Black & Associates)
	H Hemmings	Clerk of Works

Item		Action
1.0	Apologies None.	
2.0	Minutes of last meeting Agreed as correct.	
3.0	Contractor's report	
3.1	Progress is generally satisfactory.	
3.2	Still one week behind master programme due to late delivery of bricks.	
3.3	AI5 received and actioned.	
3.4	Details of ironmongery revisions required in next two weeks.	Architect
3.5	No accidents to report.	
4.0	Clerk of Works' report	
4.1	Concern expressed about poor stacking of bricks.	L&MC
5.00 – 8.00	*Continue through agenda*	
9.0	Future meetings Site Meeting – 4 November 2002 at 10.00am. Architect's site visit – 21 October 2002 at 10.00am.	All

Distribution:

Copies	2 Employer	3 Main Contractor	1 Project Manager
	1 Planning Supervisor	1 Architect	1 Quantity Surveyor
	1 Structural Engineer	1 Services Engineer	1 Clerk of Works

Example 11/3   Typical site meeting minutes.

Chapter 12
Site Duties

Supervision and inspection

It is important to distinguish between these two activities.

Supervision is the responsibility of the main contractor; he is bound by the conditions of JCT98 to complete the work in a certain time and to a specified standard. Other forms of contract may also provide or imply contractor supervision, for example a design and build arrangement.

The architect, under the terms of his employment, will normally be required to visit the site, at intervals appropriate to the stage of construction, to inspect the progress and quality of the works and to determine that they are being executed in accordance with the contract documents. While on site, it is often possible for him to make sure that problems, which are bound to arise, even on the smallest project, are satisfactorily resolved. However, the architect is not normally required to make frequent or constant inspections. The term 'inspection' has far reaching implications, inferring a depth and rigour of examination to which many architects will not wish to be committed and for which they may be open to legal recourse should any problems arise which should have been discovered during inspections. The term 'observation' is less onerous. Some bespoke forms of agreement and contract describe the architect as having to 'observe' the work on site. 'Observe' is therefore the term that we have used in conjunction with the activities of the design team throughout the remainder of this chapter.

The nature and extent of supervision arrangements will depend on the size and complexity of the works. On a small project, for example, the builder may be able to rely on a competent general foreman, while on large projects, if the job is to be properly supervised, a number of foremen and assistants may be required, working under a site agent or a contracts manager.

Similarly, the architect on a small project may be able to undertake his normal duties by periodic visits; however on large and complex projects, one or more clerks of works may be needed, with possibly a resident architect. In this chapter we are assuming that there will be a clerk of works acting pursuant to JCT 98, *'as inspector on behalf of the Employer under the directions of the Architect.'*

possibly a resident architect

The role of clerk of works may vary slightly depending on whether the appointment is made by the employer or the architect. If appointed by the employer, the role may approach that of a project manager, looking after the employer's broader interests.

Observations and inspections can be considered under the following headings:

- formal site meetings (see Chapter 11)
- routine site visits
- consultants' site visits
- inspections by statutory officials
- records and reports
- samples and testing.

Routine site visits

The architect will normally have more time to observe the work when making routine site visits than on those occasions when a formal site meeting is held. The frequency of visits and the depth of observations will depend on the size and complexity of the work and the speed of the progress being made.

As a matter of safety and courtesy, the architect should make his presence on site known to the clerk of works and the site agent. The latter will normally accompany him around the job. The architect should never give instructions to a workman directly. The main contractor, through his site agent or general foreman, is the only one with authority, under JCT98, to receive and act upon architect's instructions. Any oral instructions given during a site visit should be confirmed in writing (see Chapter 13).

When making a site visit to observe the works it is easy to be distracted and to overlook items that require attention. It is a good plan, therefore, to list in advance particular points to be looked at and any special reason for doing so. For this purpose a standard checklist is a helpful aide memoir. Provided that the contents of such a list are of a general nature, they can then be amplified or adapted for each job according to the type of work and form of construction involved. A typical standard checklist is given as Example 12/1 at the end of this chapter.

Consultants' site visits

In addition to the architect, the structural engineer and services consultants will also make regular site visits to observe the works.

Depending upon the nature of the work, it is often necessary, in the early stages of a project, for the structural engineer to confirm or adjust foundation designs following excavation and to advise on works necessary to stabilise existing buildings. Subsequently, he will be particularly interested in the testing of concrete and other structural elements (see *Samples and testing* below).

Once the services installations are under way, the services consultants are likely to undertake site visits to ensure that the works are meeting the specified requirements.

Inspections by statutory officials

Visits by statutory officials are to be anticipated on any project – with a wider range visiting the larger project.

Foremost is the building control officer from the local authority. The contractor is required to give him notice of particular critical points within the construction process, whereupon he will visit and advise/comment upon matters pertinent to the building regulations, for example, foundations, drainage, structure, fire protection, etc. It is common for an inspector to issue directions, but the contractor must, before acting upon them, refer any such direction to the architect for clarification and confirmation as an instruction. The architect may wish to consider the inspector's directions for several reasons:

- they may compromise functionality
- they may have an effect on the programme
- they may have cost implications
- there may be preferable, alternative solutions
- he may wish to discuss, with the building control officer, whether or not they are really necessary.

A local authority environmental health officer may wish to inspect the site on the grounds of noise or dust pollution.

An officer from the Health & Safety Inspectorate is likely to visit the site to inspect such things as site access, general working conditions and scaffolding to ensure that they meet relevant safety criteria. Also, he may ask to see the Risk Register.

Subject to the type of building, a Fire and Civil Defence Authority officer may visit to ensure that the means of escape are acceptable, alarms are audible, fire doors adequate, etc.

In addition to statutory officials, it may be that others, for example NHBC inspectors, adjoining owners' surveyors, or surveyors acting for insurers and the employer's financial backers, seek inspections.

Whoever visits the site, it is important that the contractor verifies the purpose of their visit and checks whether it is advisable for any other party to accompany them. The contractor should always record the visit and make a brief report of the reason for and outcome of the visit.

Records and reports

Where a clerk of works is employed he should be required to keep a daily diary of all matters affecting the project. This diary, which should be handed to the architect at the end of the project, should form the basis of the clerk of works' weekly reports to the architect. A standard form of report is available from the Institute of Clerks of Works. A completed copy is given as Example 12/2 at the end of this chapter.

Such reports cover:

- number of men employed daily, in the various trades
- weather report and particulars of time lost due to adverse weather conditions
- length of any work stoppages
- visitors to site including statutory inspectors
- delays including action taken to remedy defective work
- site directions issued
- drawings/information received on site
- drawings/information required on site
- plant/materials delivered to or removed from site
- general progress in relation to the programme
- general report including a summary of work proceeding
- comments on site conditions/cleanliness/health and safety.

... adverse weather conditions ...

The clerk of works' diary and weekly reports constitute a most useful record of events, which may be referred to should disputes subsequently arise. In the absence of a clerk of works, the contractor should be required to compile a similar factual record of conditions and activities on site and the design team should maintain their own records of site observations including plant, labour, activity and progress, to augment or amplify the contractor's records.

A copy of the contractor's programme should be kept in the office of the clerk of works and the actual progress should be checked and recorded against this programme every week.

It is the job of the clerk of works to keep records of any departures from the production information so that, on completion, the architect has all the information necessary to enable him to issue, to the employer, an accurate set of as-built drawings of the finished building. These records are particularly important where the work is to be concealed, as for example foundations, the depth of which may vary from that shown on the original production information.

Progress photographs of the work also form valuable records if taken regularly. This is probably best arranged in conjunction with the contractor so that the whole building team can have the benefit of them.

If reports are transmitted by e-mail, hard copies should be made for enduring record purposes. Those receiving reports should confirm receipt.

Samples and testing

The architect may call for samples of various components and materials used in the building to be submitted for approval, in order that he can satisfy himself that they meet the specified requirements of the employer and, where applicable, the local authority. Some of the items, of which samples would normally be required, are:

- external cladding materials such as facing bricks, artificial and natural stone, precast concrete, marble, terrazzo, slates and roofing tiles
- internal finishes such as timber, joinery, ironmongery, floor tiles and wall tiles
- services such as plumbing components, sanitary goods and electrical fittings.

In addition to samples of individual components or materials, the architect will often require sample panels to be prepared on the site to enable him to judge the effect of the materials in the situations in which they will be used. For example, a panel of facing bricks to demonstrate a particular bond or pattern, the colour of mortar and type of pointing.

It is quite normal to require laboratory tests of basic materials such as concrete. The testing of concrete should be carried out on a regular basis. Cubes should be cast from each main batch of concrete used, and each cube should be carefully labelled and identified. An approved laboratory should carry out the tests and the test reports should be submitted by the contractor to the architect or structural engineer. Full instructions for such testing procedures are usually included in the specification.

The crushing strength of bricks may also be tested in a laboratory although, unless the design requirements are stringent, a certificate from the manufacturer giving their characteristics may well suffice.

If required, manufacturers of external cladding will provide test rigs not only to investigate and demonstrate appearance and structural performance of their products, but also to test for water penetration under simulated wind pressure.

British Standard Specifications, Agrément Certificates and Codes of Practice are specified for many building materials, components and processes, and in carrying out tests it is essential to refer to the appropriate standard or code to ensure that the requirements are complied with. In addition, the British Standard Specifications set down acceptable tolerances for manufactured goods. Copies of all relevant standards and codes should always be kept on site by the clerk of works or the contractor.

Considerate Constructors Scheme

Constructing a building makes a significant impact on the quality of life in the environment around the site. The impact is particularly significant in city centres, and no more so than in the City of London. In 1987, the Corporation of London pioneered what they called the Considerate Contractors Scheme and in 1997 the Construction Industry Board launched a similar scheme, known as the Considerate Constructors Scheme. This latter scheme is a voluntary code of practice, which seeks to encourage building, demolition and civil engineering contractors to carry out their operations in a safe and considerate manner. The Considerate Constructors Scheme, which is operated by the Construction Confederation, has now been adopted across the country. Annual awards are made under the scheme to recognise and reward the contractors' commitment to raise standards of site management, safety and environmental awareness beyond their statutory duties. The Code of Practice for the Considerate Constructors Scheme includes the following criteria:

- **Considerate** – all activities are to be carried out with positive consideration for the needs of traders and businesses, site personnel and visitors, and the general public. Special attention is to be given to the needs of those with disabilities.
- **Environment** – noise from construction work, site personnel and other sources is to be kept to a minimum at all times. Local resources should be used wherever possible. Waste management and the avoidance of pollution by recycling surplus materials are to be encouraged.
- **Cleanliness** – the working site, safety barriers, lights and signs are to be maintained in a clean and safe condition. Surplus materials and rubbish must not be allowed to accumulate on site or spill over onto adjacent property. Dust from all activities is to be kept to a minimum.
- **Good neighbours** – full and regular consultation with neighbours regarding programming and site activities is to be maintained from pre-start to completion of activities.
- **Respectful** – all site personnel are to wear respectable and safe dress, appropriate to the weather conditions and are to be instructed in dealing with the general public. Lewd or derogatory behaviour and language are not to be tolerated. The contractor is to take pride in the management and appearance of the site and surrounding environment.
- **Safe** – movements of site personnel and vehicles are to be carried out with great care and consideration for traders and businesses, site personnel and visitors, and the general public. No building activity is to be a security risk to others.
- **Responsible** – the contractor is to ensure that all persons working on the site understand and implement the Code of Practice.

- **Accountable** – posters are to be displayed clearly around the site, giving names and telephone numbers of staff who can be contacted to respond to issues raised by those affected.

Site safety

Construction sites should be safe places for those who work on them, and for those who are neighbours. The route to site safety is through the Safety Plan, which falls under CDM94 and the Health and Safety at Work Act 1974 (HSWA).

Section 2(3) of the HSWA, states:

> *'Except in such cases as may be prescribed, it shall be the duty of every em-ployer to prepare and as often as may be appropriate revise a written state-ment of his general policy with respect to the health and safety at work of his employees and the organisation and arrangements for the time being in force for carrying out that policy, and to bring the statement and any revision of it to the notice of all his employees.'*

All employers who employ five or more employees at any one time must pre-pare such a statement, the main purpose of which is to identify risks involved at work, to identify the precautions and to show who is responsible for carry-ing out those precautions. The essential ingredients of a good policy statement are:

- a statement of the employer's general policy with regard to health and safety
- details of the organisational/managerial hierarchy for carrying out the policy
- details of the practical arrangements for carrying out the policy
- the name of the person in authority who is responsible for fulfilling the policy.

Although each company is responsible for drafting its own policy statement according to its own needs, a model policy document is available from HSE Books.

The law only requires safety policy statements to cover the health and safety of employees. However, the contractor should state his strategy for protecting other people who could be put at risk by his activities, such as sub-contrac-tors, customers and the public. If the activities of others on site, such as sub-contractors, could put employees at risk, the contractor will need to consider how these risks are to be avoided and to cover this aspect in his statement.

If the contractor has a safety committee, its constitution and terms of refer-ence should be clearly established. Further information on safety committees can be found in the Safety Representatives and Safety Committees Regulations 1977 SI 1977 No 500 and Safety Representatives and Safety Committees booklet (ISBN 0 7176 0419 5), which contains the regulations.

Issues to be considered by the contractor when drafting his safety policy statement are many and various and include:

Generally:
- The statement should express a commitment to health and safety and make clear the contractor's obligations towards their employees.
- The statement should record which person in authority is responsible for ensuring that it is implemented and regularly reviewed.
- The statement should be signed and dated by a partner or senior director.
- The statement should take account of the views of managers, supervisors, safety representatives and the safety committee.
- The duties set out in the statement should be discussed with the people concerned in advance and accepted by them, to ensure that they understand how their performance is to be assessed and what resources they have at their disposal.
- The statement should make clear that co-operation on the part of all employees is vital to the success of the contractor's health and safety policy.
- The statement should record how employees are to be involved in health and safety matters, for example by being consulted, by taking part in inspections and by sitting on a safety committee.
- The statement should show clearly how the duties for health and safety are allocated and describe the responsibilities at different levels.
- The statement should record who is responsible for the following matters (including deputies where appropriate):
  - reporting investigations and recording accidents
  - fire precautions, fire drill, evacuation procedures
  - first aid
  - safety inspections
  - the training programme
  - ensuring that legal requirements are met, for example regular testing of lifts
  - notifying accidents to the health and safety inspector.

Site safety:
- Delivery access, timing and loading restrictions.
- Storage compound location.
- Inflammable product storage.
- De-commissioning, storage or disposal of any existing equipment.
- Position of tool store, welfare facilities and shared facilities.
- Security and control of the access to high-risk work areas, machinery rooms, control equipment and high voltage switch gear.

- Scaffolding and protection arrangements.
- Permits-to-work for flame producing or oxy-acetylene cutting gear.
- Control of Substances Hazardous to Health (COSHH) assessments of dusty works or substances used in occupied or shared work areas.
- Temporary lighting requirements or specification of power locations and usage.
- Risk assessments of work operations, relevant in detail and scale, to the project in progress.
- Method statements relating to the implementation of high risk elements of the work, giving details and descriptions relevant to the size and scale of the works.
- Commissioning logs and partial hand-over arrangements.
- Lift usage restrictions/authorised user list.
- Arrangements for dealing with fire prevention.
- Training and competence certificates.
- First aid arrangements.
- Emergency contact list.
- Named management and responsibilities.
- Details of how the archived information will relate to the safety file and what operation and maintenance information will be given to the principal contractor during the project.
- Keeping the workplace, including staircases, floors, ways in and out, washrooms etc in a safe and clean condition by cleaning, maintenance and repair.

Plant and substances:
- Maintenance of equipment such as tools and ladders in a safe condition.
- Maintenance and proper use of safety equipment such as helmets, boots, goggles, respirators, etc.
- Maintenance and proper use of plant, machinery and guards.
- Regular testing, maintenance and emergency repair of lifts, hoists, cranes, pressure systems, boilers and other dangerous machinery.
- Maintenance of electrical installations and equipment.
- Safe storage, handling and, where applicable, packaging, labelling and transport of dangerous substances.
- Controls on work involving harmful substances such as lead and asbestos.
- The introduction of new plant, equipment or substances into the workplace – by examination, testing and consultation with the workforce.

Other hazards:
- Wearing of ear protection, and control of noise at source.
- Preventing unnecessary or unauthorised entry into hazardous areas.
- Lifting of heavy or awkward loads.

- Protecting the safety of site staff against assault when handling or transporting the employer's money or valuables.
- Special hazards to site staff when working on unfamiliar sites, including discussion with site manager where necessary.
- Control of works transport, such as forklift trucks, by restricting use to properly trained, experienced operatives.

Emergency procedures:
- For ensuring that fire exits are marked, unlocked and free from obstruction.
- For maintenance and testing of fire-fighting equipment, fire drills and evacuation.
- Names and location(s) of person responsible for first aid and their deputy and the location of the first aid box.

Communication procedures:
- For giving site staff information about their general duties under the HSWA and specific legal requirements relating to their work.
- For giving site staff necessary information about substances, plant, machinery and equipment with which they come into contact.
- For discussing with sub-contractors, before they come on site, how they can plan to do their job, whether they need contractor's equipment to help them and what hazards they may create for site staff and vice versa.

Training procedures:
- For employees, supervisors and managers to enable them to work safely and to carry out their health and safety responsibilities efficiently.

Supervision procedures:
- For site staff so far as is necessary for their safety – especially young workers, new employees and employees carrying out unfamiliar tasks.

Checking procedures:
- For making regular inspections and checks of the workplace, machinery, appliances and working methods.

Fire precautions on site

Fire is a serious and ever-present danger on site and is a key issue when considering site safety. Fires spread rapidly and can endanger life and cause damage to property, including collateral damage to neighbouring buildings and equipment – there are about 11 construction site fires each day. Fire precautions depend on the nature and location of the project. Erecting a simple steel-framed building in a rural location will require only simple precautions because fire risks are minimal. Higher risk work, for example refurbishing floors in an occupied office block, will need many more precautions because the risk of fire occurring, the difficulties of escape and the potential for injury and loss of life are that much greater. In any event, fire alarm points, local extinguishers (to enable escape, not to fight the fire) must be available to comply with regulations throughout the site.

Preventing fires on site is approached in three principal ways:

- First, the Construction (Health, Safety and Welfare) Regulations 1996 require contractors to take measures both to prevent fires occurring and to ensure that everyone on construction sites, including visitors, is protected if fires happen.
- Second, the Construction (Design and Management) Regulations 1994 (CDM) require those designing, planning and constructing projects to take construction fire safety into their thinking.
- Third and arguably most important, The Joint Code of Practice on the Protection from Fire of Construction Sites and Buildings Undergoing Renovation (Joint Fire Code) published by the Construction Confederation and the Fire Protection Association.

The Joint Fire Code

This Code applies to works of £1m in value and over. Works of £25m in value and over are termed 'Large Projects' for which the Code provides additional requirements. In the fifth edition of the Joint Fire Code, which was published in January 2000, it states in clause 2 that

> *'Non-compliance with the Code by the Construction Industry, by those who procure construction and by construction industry professionals could result in insurance ceasing to be available or being withdrawn resulting in a possible breach of construction contracts which require the provision of such insurance.'*

Clause 22FC of JCT98 provides for compliance with the Joint Fire Code by both the employer and the contractor when it applies to their particular contract. It also makes provisions to ensure and that any 'Remedial Measures' required by the insurer under the Joint Names Policy, to rectify a breach of the Joint Fire Code, are carried out timeously.

An example of the code's requirements is the use on site of high-quality flame retardant protection materials approved to Loss Prevention Standard LPS 1207.

Preventing fire

Basic precautions to prevent fires on construction sites are as follows:

- Store LPG cylinders and other flammable materials properly. LPG cylinders should be stored outside buildings in well ventilated and secure areas. Other flammable materials, such as solvents and adhesives should be stored in lockable steel containers. Care must be taken to avoid leaks.
- Carefully control hot work such as welding by using formal permit-to-work systems.
- Do not leave tar boilers unattended.
- Keep an orderly site and make sure rubbish is cleared away regularly.
- Avoid unnecessary stockpiling of combustible materials such as polystyrene.
- Take special precautions in areas where flammable atmospheres may develop, for example when using volatile solvents or adhesives in enclosed areas.
- Avoid burning waste materials on site wherever possible. **Never** use petrol or similar accelerants to start or encourage fires.
- Make sure everyone obeys site rules on smoking.

Means of escape

Construction sites can pose particular problems because the routes in and out may be incomplete and obstructions may be present. Open sites usually offer plentiful means of escape and special arrangements are unlikely to be necessary. In enclosed buildings people can easily become trapped, especially where they are working above or below ground level. In such cases means of escape need careful consideration. The contractor must make sure that:

- Wherever possible, there are at least two escape routes in different directions.
- Travel distances to safety are reduced to a minimum.
- Enclosed escape routes, for example corridors or stairwells, can resist fire and smoke ingress from the surrounding site. Where fire doors are needed for this, the contractor must ensure they are provided and kept closed; self-closing devices should be fitted to doors on enclosed escape routes.
- Escape routes and emergency exits are clearly signed.
- Escape routes and exits are kept clear. Emergency exits should **never** be locked when people are on the site.
- Emergency lighting is installed if necessary to facilitate escape. This is especially important in enclosed stairways in multi-storey structures, which will be in total darkness if the normal lighting fails during a fire.
- An assembly point is identified where everyone can gather and be accounted for.

Fire-fighting equipment

The equipment needed depends on the risk of fire occurring and the likely consequences if it does. It can range from a single extinguisher on small low-risk sites to complex fixed installations on large and high-risk sites. In any event the contractor must ensure that:

- Fire-fighting equipment is located where it is really needed and is easily accessible.
- The location of fire-fighting equipment and instructions on its use are clearly indicated.
- The right sorts of extinguishers are provided for the type of fire that could occur. A combination of water or foam extinguishers for paper and wood fires and CO_2 extinguishers for fires involving electrical equipment is usually appropriate.
- The equipment provided is maintained and works. A competent person, normally from the manufacturer, should check fire-fighting equipment regularly.
- Those carrying out hot work have appropriate fire extinguishers with them and know how to use them.

Emergency plans

The purpose of emergency plans is to ensure that everyone on site reaches safety if there is a fire. Small and low-risk sites only require very simple plans, but higher risk sites will need more careful and detailed consideration. An emergency plan should:

- be available before work starts on site
- be up to date and appropriate for the circumstances concerned
- make clear who does what during a fire
- be incorporated in the construction phase health and safety plan when CDM94 applies
- work if it is ever needed.

Providing information

Fire action notices should be clearly displayed where everyone on site will see them, for example at fire points, site entrances or canteen areas.

This list should be amplified or adapted according to the nature of the project and may serve the architect or engineer. Some of the items should be observed jointly with other consultants.

General:
- In all cases check that the work complies with the latest drawings and specification, with the latest requirements of the statutory undertakers and with the building regulations. Ensure that all information is complete.

Preliminary works:
- scaffolding and the Building Act 1984
- location of workmen's canteens, builder's offices, etc.
- suitability and location of clerk of works' or site architect's office
- removal of top soil and location of spoil heaps
- perimeter fencing or hoardings
- protection of rights of way
- protection of trees and other special site features
- protection of materials on site
- party-wall agreements and protection of adjoining property
- site security generally
- agreement on bench mark or level pegs
- agreement on setting out (responsibility of the contractor).

Demolition:
- extent
- adequacy of shoring
- preservation of certain materials and special items (to be listed in the specification or bills of quantities).

Excavation and foundations:
- widths of trenches
- depths of excavations
- nature of ground in relation to trial hole report
- stability of excavations
- pumping arrangements
- risk to adjoining property and general public
- quality of concrete and thickness of beds
- suitability of hardcore (freedom from rubbish)
- suitability of sand and ballast (freedom from loam and correct grading)
- damp-proof membranes and tanking
- ducts, drains or services under building
- size, bending, spacing and placing of reinforcement
- suitability of material for and consolidation of backfilling
- depths of piles and driving/boring conditions.

Drainage:
- depths of inverts and gradients of falls
- thickness and type of bed and jointing of pipes
- quality of bricks for manholes and rendering thereto
- testing of drains and manholes.

Brickwork, blockwork and concrete masonry:
- approval of sample panels of facings and fairface work
- quality and colour of mortar and pointing
- test report on crushing strength where necessary
- BS certificates on load-bearing blocks
- position and type of wall ties

Example 12/1   Standard checklist for site inspections. (*Continued.*)

- cleanliness of cavities and wall ties in external walls
- setting-out and maintenance of regular vertical and horizontal joints
- type, quality and placing of damp proof courses
- setting-out and fixing of door frames, windows, etc.
- bedding and levels of lintels over openings
- expansion joints.

In-situ concrete:
- setting out and stability of shuttering
- shuttering type to achieve finish specified
- setting out of reinforcement, fixings, holes and water bars
- mix for and procedure of taking test cubes
- curing of concrete and striking of shuttering
- vibration
- use of additives.

Precast concrete:
- size and shape of units
- quality of fit and finish
- position of fixings, holes, etc.
- damage in transit and erection.

Carpentry and joinery:
- freedom from loose knots, shakes, sapwood, insect attack, etc.
- dimensions within permissible tolerances
- application of timber preservatives and primers
- storage, stacking and protection from weather
- jointing, bolting, spiking and notching of carpenter's timber
- spacing of floor and ceiling joists and position of trimmers
- spacing of battens, position of noggings
- weather throatings and cills to doors, windows, etc.
- jointing, machining and finish of manufactured joinery
- fire rating.

Roofing:
- pitch of roof
- spacing of rafters and tile battens
- approval of under felt
- approval of roofing materials and fixings
- pointing to verges and bedding of ridges, etc.
- falls to outlets on flat roofs
- thickness and fixing of insulation under finish
- eaves details and ventilation
- correct formation of flashings, etc.
- ventilation.

Cladding:
- vapour barriers and insulation
- regularity of grounds for sheet materials
- location and quality of fixings
- laps, tolerances, and positions of joints
- setting-out and jointing of mullions and rails
- handling and protection of panel materials
- specification and application of mastic
- flashings, edge trims and weather drips

Example 12/1   (*Continued.*)

- entry and egress of moisture
- prevention of electrolytic action
- location and type of movement joints.

Steelwork:
- sizes of steel
- rivets and welding
- position of members
- plumbing, squaring and levelling of steel frame
- priming and protection.

Metalwork:
- sizing and spacing of members
- galvanising and rustproofing
- stability of supports, including caulking or plugging
- isolation from corrosive materials, etc.

Plumbing and sanitary goods:
- ensuring that sanitary goods are free from cracks and deformities
- location, venting and fixing of stack pipes
- falls to waste branches
- use of traps
- jointing of pipes
- smoke and/or water tests
- location and accessibility of valves, stop-cocks
- drain-down cocks at lowest points
- access to traps and rodding eyes.

Heating, hot water and ventilation installations:
- types of boiler, cylinder, tanks, fans, etc.
- types of pipes
- position and type of stop valves
- position of pipe runs and ventilation ducting
- insulation of pipework and ducting
- identification and labelling of pipes, valves, etc.

Electrical installation:
- types of switchgear, distribution board, motor drives, etc.
- types of switches, socket outlets, fuses, cables, etc.
- location of outlet points
- earthing of installation
- lightning conductor installation
- runs of cables and quality of connections, etc.
- labelling and identification of switchgear, etc.

Specialist installations:
- drawings of specialist installations
- drawings of builder's work in connection with specialist installations
- power and plant to be provided
- access for equipment and provision of adequate working space
- temporary support, loading on structure, lifting tackle, etc.
- attendance on site and sequence of work.

Paving and floor tiling:
- approve materials
- quality of screeds to receive flooring
- junctions of differing floor finishes

Example 12/1   (*Continued.*)

- regularity of finish
- falls to gulleys, etc.
- skirtings and coves
- expansion joints
- types of tile bedding and grouting materials.

Plastering and wall tiling:
- storage of materials
- plaster mixes
- preparation of surface
- true surfaces and arrises
- fixing of plasterboard
- filling and scrimming of joints in plasterboard
- adequate hacking of or bonding plaster on concrete
- regularity of finish
- types of tile bedding and grouting materials.

Suspended ceilings:
- type of suspension and tile
- height of ceiling and setting out
- location and co-ordination of light fittings, sprinkler outlets, ventilation grilles, etc.
- access panels
- finish trim for curtains, blinds, etc.
- fire barriers.

Glazing:
- quality of glass and freedom from defects
- integrity of sealed units
- structural capability
- type and/or thickness
- depth of rebates
- glazing compound, fixing of glazing beads, etc.

Painting and decorating:
- approval of materials
- preparation of surfaces and freedom from damp
- ensuring that partly concealed surfaces are properly finished
- ensuring that finished work is free from runs, brush marks, etc.
- opacity of finish, etc.

Cleaning down and handing over:
- windows cleaned and floors scrubbed
- sanitary goods washed and flushed
- painted surfaces immaculate
- doors correctly fitted, windows not binding or rattling
- ironmongery complete and locks and latches operating correctly
- correct number of and suiting of keys and security cards
- connection of services, provision of meters, etc.
- commissioning of all mechanical engineering plant, balancing of air-conditioning, etc.
- plant maintenance manuals, plant room service diagrams, etc.
- operation of security, communication and fire protection systems
- removal of protective tapes and films etc.
- cleaning or replacement of air filters in the HVAC system
- cleaning of lighting reflectors where uncovered.

Example 12/1   (*Continued.*)

INSTITUTE OF CLERKS OF WORKS

CLERK OF WORKS PROJECT REPORT

NO ....6....

PROJECT: ....Shops & offices.... REF: ....

ADDRESS: ....Newbridge Street, Borchester....

Architect/Contract Administrator ....Reed & Seymore....

Main Contractor ....Leavesden Barnes....

Clerk of Works Phone/Fax No ....

Week/Month Ending ....14 December 2001....

Contract Start Date ....2 November 2001....

Contract Completion Date ....13 September 2002....

Progress +/- to Programme ....

TRADES	Mon	Tues	Wed	Thur	Fri	Sat	Sun	Delays including Defective Work (Action Taken)
Site Staff	3	3	2	3	3			3m length of drainage trench collapsed
Groundworkers	2	2	2	2	2			(NE corner) 12 Dec. Re-excavation required
Steelfixers								with sheet piling. 1 Day lost.
Steel Erectors	2	2	2					
Concretors								
Drainlayers	1	1	1	1	1			
Machine Operators								
Carpenters	2	2	2	1	2			Site Directions Issued
Scaffolders								
Bricklayers	4	4	4	2	2			
Roof Finishers								
Wall Cladding								
Window Fixers								
Glaziers								Drawings/Information Received on Site
Floor Screeders								Roof details - dwg nos. 456/55A, 56B, 57
Plasterers								
Tilers-Wall/Floor								
Dryliners/Partitions								
Ceiling Fixers								Drawings/Information Required on Site
Decorators								Finishes details
Floor Finishers								
Heating/Ventilation								
Plumbing								
Electricians								Plant/Materials Delivered to Site or Removed
Hard/Soft Landscape								Steelwork delivered
Roadworks								
Public Services								
Site agent	1	1		1	1			
								General Comments
TOTAL	15	15	13	10	11			Steelwork primer badly scratched

Site Directions Issued

No	Item	Date
4	Manhole 3 repositioned	10 Dec

Contractors Labour Returns May Be Substituted

WEATHER REPORT					Time Lost
	AM	°C	PM	°C	
Mon	Fine	5	Fine	6	
Tues	"	4	"	5	
Wed	overcast	4	overcast	4	
Thur	heavy rain	4	rain	3	1 day
Fri	Fine	3	Fine	3	
Sat					
Sun					

Stoppages (Hours)

Total to Date

Visitors (Include Statutory Inspectors)	Date
Building Inspector	10 Dec
Project architect	12 Dec
Health & Safety Inspector	12 Dec

Example 12/2 Clerk of works project report. (*Continued.*)

	Progress to date %	Programme %		Progress to date %	Programme %		Progress to date %	Programme %
Preliminaries			Blockwork Internal		·	Electrical 2nd Fix		
Excavation	80	85	Cladding/Curtain Wall			H & V 2nd Fix		
Shutter/Reinf			Windows-glazing			Ceiling Grid/Tiles		
Concrete Structure			Joinery 1st Fix			Decoration		
Steel Erection	15	15	Plastering		·	External Works		
Main Drainage m/h	50	60	Drylining/Partitions			Hard/Soft Landscape		
Floor Construction			Floor Screeds			Roadworks		
Floors Suspended			Plumbing 1st Fix			Mains; Gas-Electrical		
Roof Structure			Electrical 1st Fix			Mains; Water-Telecoms		
Roof Coverings			H & V 1st Fix			Lifts		
Drainage fw. sw.	25	30	Wall/Floor Tiles			Alarm/Computer Systems		
Brickwork External			Plumbing 2nd Fix			Defects/Handover		

GENERAL REPORT: Summary of Work Proceeding

Excavations largely completed
Drainage progressing well — but see note under "delays"
Building inspector satisfied with footings
Steelwork started this week

Site Conditions/Cleanliness/Health & Safety: Action Taken

Conditions generally good due to recent fine weather
Reason for collapse of drainage trench being investigated

Enclosures: G.C. Labour Return [ ] Site Directions [1] Reports (State) [ ]	OFFICE ACTION:
Distribution as Agreed	
Client [1] Architect [1] Project Manager [·] Quantity Surveyor [1] Office [ ] Others [ ]	

Clerk of Works   Harry Hemmings   Date   17 December 2001

Example 12/2   *(Continued.)*

Chapter 13

Instructions, Variations and Post-Contract Cost Control

Provided that the procedures set out in the preceding chapters are followed, it should be possible to have available complete sets of drawings, specification notes and nominated sub-contractors' and suppliers' quotations when the tender documents are prepared. This in turn will mean that, as soon as the contract is placed, the contractor can be handed all the necessary information to enable him to build the project.

Architect's instructions

The ideal circumstances outlined above are not the norm, however, and even when fully finalised information is available for incorporation into the contract documents, it will still be necessary, from time to time, for the architect to issue further drawings, details and instructions. These are collectively known as architect's instructions. The JCT98 conditions of contract set out those matters in connection with which the architect is empowered to issue such instructions and these are as follows:

Clause	Description
2.3 & 2.4.1	Discrepancies in or divergences between documents
5.4.2	Provision of further drawings or details
6.1.3	Statutory Requirements
6.1.6	Divergence – Statutory Requirements and the Contractor's Statement
6.1.7	Change in Statutory Requirements after Base Date
7	Levels and setting out of the Works
8.3	Inspection – tests
8.4	Powers of Architect – work not in accordance with the Contract
8.5	Powers of the Architect – non-compliance with clause 8.1.3
8.6	Exclusion from the Works of persons employed thereon
13.2.1	Instructions requiring a variation

13.3.1 & 13.3.2   Instructions on provisional sums
13A.4   Contractors 13A Quotation not accepted
17.2   Defects, shrinkages and other faults
17.3   Defects, etc. – Architect's instructions
21.2.1   Insurance – liability, etc. of Employer
22D.1   Insurance for Employer's loss of liquidated damages
22FC.3.1.2   Breach of Joint Fire Code – Remedial Measures
23.2   Architect's instructions – postponement
30.6.1.1   Final adjustment of Contract Sum – documents from Contractor
34.2   Architect's instructions on antiquities found
35.5.2   Architect's instruction on NSC/N – documents accompanying the instruction
35.6   Architect's instruction on Nomination NSC/N – documents accompanying the instruction
35.9.2   Architect's duty on receipt of any notice under clause 35.8
35.18.1.1   Defects in nominated sub-contract works after final payment of Nominated Sub-Contractor – before issue of Final Certificate
35.24.6, 7 & 8   Circumstances where re-nomination necessary
35.25   Determination of employment of Nominated Sub-Contractor – Architect's instructions
36.2   Nominated Suppliers – Architect's instructions
42.14   Integration of Performance Specified Work

The procedure for the issue of architect's instructions is set out in JCT98 clause 4.3 and can be summarised as follows:

- Any instruction issued by the architect must be in writing, but
- should the architect issue an instruction other than in writing it is of no effect unless, within seven days of such an instruction being given:
  - it is confirmed in writing by the architect, at which time it becomes effective, or
  - it is confirmed in writing by the contractor and is not dissented from by the architect within a further period of seven days, at which time it becomes effective; however
- if neither the architect nor the contractor confirms an instruction issued other than in writing but the contractor complies with the instruction, the architect can, at any time up to the issue of the final certificate, confirm that instruction in writing, at which time the instruction becomes effective.

Whilst it is worth noting that JCT98 clause 4.2 allows the contractor to question the contractual validity of any architect's instruction and requires the architect to answer 'forthwith', it should also be noted that, if the contractor does not comply with a valid architect's instruction, clause 4.1.2 allows the employer to

Any instruction issued by the architect must be in writing

employ others to give effect to the instruction and recover all costs so incurred from the defaulting contractor.

If a clerk of works is employed on the project and issues any directions to the contractor, such directions are effective only if they are issued in regard to a matter in respect of which the architect is empowered to issue instructions, and also if they are confirmed in writing by the architect within two working days (not, it will be noted, within seven days as is provided for the confirmation by the architect of his own instructions issued other than in writing). Standard forms are available for such clerk of works' directions.

It is good practice for all instructions from the architect to the contractor to be issued, or confirmed, on standard forms. An example of such a form, which is published by RIBA Enterprises Ltd, is given as Example 13/1 at the end of this chapter.

It is essential that instructions should be clear and precise and, where revised drawings are issued, the date and reference of the particular revision should be specifically referred to. Instructions emanating from other members of the design team must not be given directly to the contractor but must be issued to him via the architect as architect's instructions. Copies of architect's instructions should be distributed to:

- contractor
- employer or project manager or employer's representative
- planning supervisor
- quantity surveyor
- other members of the design team
- clerk of works
- any nominated sub-contractor affected by the instruction.

Architect's instructions nominating sub-contractors and suppliers are covered by JCT98 clauses 35.6 and 36.2 respectively and are dealt with in more detail in Chapter 8. It should be noted that JCT98 clause 13.1.3 precludes the architect from issuing an instruction nominating a sub-contractor to carry out work which has been measured in the bills of quantities and priced by the contractor for execution as main contractor's work. If the architect wishes to do this, he may do so only with the contractor's agreement. If the contractor does agree it would be wise to settle the arrangements for his profit, attendance on the sub-contractor and any other financial matters before the nomination is made.

JCT98 also makes provision for the architect to issue instructions to the quantity surveyor and these are as follows:

Clause	Description
26.1, 26.4.1 & 34.3.1	To ascertain the amount of loss and/or expense incurred by the contractor.
30.5.2.1	To prepare a statement specifying the details of the retention deducted in arriving at the amount stated as due in interim certificates.

Variations

It is important not to confuse architect's instructions with variations. Whilst all variations arise as a consequence of architect's instructions, not all architect's instructions are variations. In JCT98, clause 13 provides the definition of a variation as follows:

'13.1   *The term 'Variation' as used in the Conditions means:*
13.1.1   *the alteration or modification of the design, quality or quantity of the Works, including*
13.1.1.1 *the addition, omission or substitution of any work,*
13.1.1.2 *the alteration of the kind or standard of any of the materials or goods to be used in the Works,*
13.1.1.3 *the removal from the site of any work executed or materials or goods brought thereon by the Contractor for the purpose of the Works other than work, materials or goods which are not in accordance with the Contract;*

13.1.2   *the imposition by the Employer of any obligations or restrictions in regard to the matters set out in clauses 13.1.2.1 to 13.1.2.4 or the addition to or alteration or omission of any such obligations or restrictions so imposed or imposed by the Employer in the Contract Bills in regard to:*

13.1.2.1 *access to the site or use of any specific parts of the site;*

13.1.2.2 *limitations of working space;*

13.1.2.3 *limitations of working hours;*

13.1.2.4 *the execution or completion of the work in any specific order, but excludes:*

13.1.3   *nomination of a sub-contractor to supply and fix materials or goods or to execute work of which the measured quantities have been set out and priced by the Contractor in the Contract Bills for supply and fixing or execution by the Contractor.'*

The matters set out in JCT98 clause 13.1.2 relate to the working conditions imposed by the employer, particulars of which will have been given in the bills of quantities in accordance with Section A of SMM7R, or in the specification. It should be noted that whilst there is a general obligation, under JCT98 clause 4.1.1, for the contractor to comply forthwith with all architect's instructions, clause 4.1.1.1 allows that the contractor need not comply with an architect's instruction requiring a variation within the meaning of clause 13.1.2 so long as he has given the architect reasonable, written objection to such compliance. The reason for giving the contractor this important right of objection may well be in recognition of the possibility that such a variation could so fundamentally affect the basis on which the contractor tendered that the valuation of variations and direct loss and/or expense provisions within JCT98 would not provide adequate recompense for the contractor.

Valuing variations

All variations, other than those for which a contractor's price statement has been accepted under the terms of JCT98 clause 13.4 Alternative A or for which a contractor's 13A quotation has been accepted under the terms of JCT98 clause 13A, fall to be valued by the quantity surveyor in accordance with the provisions of JCT98 clause 13.5. This clause prescribes, in effect, five basic methods for the valuation of all variations, all work executed by the contractor as the result of architect's instructions as to the expenditure of provisional sums and the execution of work for which an approximate quantity is included in the contract bills.

    The method of valuation to be adopted by the quantity surveyor depends upon the nature of the work and the conditions under which it has to be executed. Each basic method of valuation and the nature of the work to which it is to be applied are summarised in Table 13.1.

Table 13.1 The five basic methods for the valuation of all variations.

Description of work	Method of valuation
1. Additional or substituted work that can properly be valued by measurement and where the work is of similar character to, is executed under similar conditions as, and does not significantly change the quantity of work set out in the contract bills. Work for which an approximate quantity is included in the contract bills and that quantity is reasonably accurate. Omission of work set out in the contract bills.	Measure in accordance with SMM7R. Value at the rates and prices for the work set out in the contract bills. Allow for any percentage or lump sum adjustments in the contract bills. Allow for any addition to or reduction of preliminary items of the type referred to in SMM7R section A except in the case of work resulting from an instruction as to the expenditure of a provisional sum in the contract bills for defined work.
2. Additional or substituted work that can properly be valued by measurement and where the work is of similar character to work set out in the contract bills but is not executed under similar conditions thereto and/or significantly changes the quantity thereof. Work for which an approximate quantity is included in the contract bills and that quantity is not reasonably accurate. Any work executed under changed conditions as a result of a variation, compliance with an instruction as to the expenditure of a provisional sum or the execution of work for which an approximate quantity is included in the contract bills.	As 1 above plus a fair allowance for the differences arising from the work not being executed under similar conditions to and/or because the work changes the quantity of the work of similar character in the contract bills.
3. Additional or substituted work that can properly be valued by measurement and where the work is not of similar character to work set out in the contract bills.	Measure in accordance with SMM7R. Value at fair rates and prices. Allow for any percentage or lump sum adjustments in the contract bills. Allow for any addition to or reduction of preliminary items of the type referred to in SMM7R section A except in the case of work resulting from an instruction as to the expenditure of a provisional sum in the contract bills for defined work.
4. Additional or substituted work that cannot properly be valued by measurement.	Provided that vouchers specifying the time spent upon the work, the workmen's names, the plant and the materials employed are delivered to the architect for verification not later than the end of the week following that in which the work is executed, value at the prime cost of such work calculated in accordance with the 'Definition of Prime Cost of Daywork carried out under a Building Contract' issued by the Royal Institution of Chartered Surveyors and the Construction Confederation or the Electrical Contractors' Association or the Heating and Ventilating Contractors' Association, as appropriate, and current at the base date together with percentage additions to each section of the prime cost at the rates set out in the contract bills.
5. Work other than additional, substituted or omitted work. Work or liabilities directly associated with a variation that cannot reasonably be valued by any other method.	A fair valuation.

... the work is of similar character ...

JCT98 clause 13.5.6 provides separate rules for the valuation of perform-ance-specified work and any variations required to it. Whilst these rules dif-fer from the variation valuation rules in detail, they respect the same basic principles.

It must be emphasised that no allowance is to be made in any valuation carried out under JCT98 clause 13.5 for any effect upon the regular progress of the works or for any other direct loss and/or expense for which the contrac-tor would be reimbursed by payment under any other provision of JCT98, for example, under clause 26 – loss and expense caused by matters materially affecting regular progress of the works.

The valuation of variations to nominated sub-contract works, including the valuation of work carried out against provisional sums included in the sub-contract, is to be made in accordance with the relevant provisions of the sub-contract. The valuation rules in the sub-contract conditions are similar in principle to those of the main contract referred to above, and we do not propose to deal with them further here. In this connection, however, JCT98 does clarify the position where the main contractor tenders successfully for nominated sub-contract works. Such work has to be valued in accordance with the accepted tender of the contractor and is not included in the valuation of the architect's instruction in regard to the expenditure of the provisional sum giving rise to the work.

It is interesting to note that when the contract rules for valuing variations are brought into play, the quantity surveyor has a unilateral responsibility – the contractor is not involved, apart from being entitled to be present when any measurements are made. Should the contractor not be satisfied with the quantity surveyor's valuation, his only formal recourse is to use the dispute resolution procedures set out in the contract. Practically, of course, the quantity surveyor usually works closely with the contractor's surveyor so that (more often than not) disputes are avoided and an agreed final account is produced.

Dayworks

Works valued at prime cost on a daywork basis effectively involve a cost-plus method of reimbursement and, from a contractor's viewpoint, this is therefore an attractive means of securing payment for variations. For work to be valued at prime cost on a daywork basis the record sheets must be submitted to the architect or his authorised representative, usually the clerk of works, no later than one week following the week in which the work was carried out. In addition to being serially numbered it is essential that the following information be recorded on each sheet:

- the reference of the architect's instruction authorising the work
- the date(s) when the work was carried out
- the daily time spent on the work, set against each operative's name
- the plant used
- the materials used.

The architect or his authorised representative should, as soon as possible after receiving the sheets, check the accuracy of the records and, if found to be correct, countersign them and pass them to the quantity surveyor who will determine whether or not the recorded work is to be valued at prime cost. It is worth noting that the architect's signature on a daywork sheet does not constitute a variation, nor does it commit the quantity surveyor to having to value the work at prime cost.

It is helpful if the contractor gives the architect advance warning of his intention of recording anything as daywork so that the architect, or the clerk of works on his behalf, can arrange for particular notice to be taken of the resources used.

Cost control

Cost control may be defined as the controlling measures necessary to ensure that the authorised maximum cost of the project is not exceeded. It is a continuous process and follows on from the pre-contract cost control activities discussed in Chapter 4.

It is essential for the design team to establish, at the outset, the parameters within which the employer requires the construction costs to be controlled. Usually employers have a limit on the amount of money available for expenditure on a project and will insist that their limit is not exceeded. It has to be said, however, that occasionally the need to achieve a high-quality end product is more important to the employer than cost.

Initially, the authorised expenditure limit for a project will more often than not be the contract sum. However, it is not uncommon for the limit to be varied during construction – for example, an employer, building speculatively, may find that a prospective tenant is prepared to pay more for an enhanced specification, in which case the employer will require the design team to assess the cost and time implications of incorporating the enhanced specification into the project and, if economically viable, the employer will approve an increase in the previously authorised expenditure limit.

As most construction projects are complex it is normal for a contingency sum to be included in the contract sum to cater for expenditure on unforeseen items of work that become necessary during the construction process. The amount of the contingency sum will vary according to the nature of the project and the perceived risk of expenditure on unforeseen items. A refurbishment project, for example, would most likely require a larger contingency provision than would a similar value new building project on a green-field site. It would be normal for the contingency sum to equal between 3% and 5% of the contract sum, although it could be more on a particularly risky project. The primary purpose of the contingency sum is to fund additional work that could not reasonably have been foreseen at design stage, for example additional work below ground; it is not there to fund design alterations, except with the prior approval of the employer. In circumstances where no unforeseen items of work arise, the contingency sum should remain unspent at practical completion but, in practice, this is a rare (indeed almost unheard of) occurrence.

If construction costs are to be controlled successfully, it is essential for there to be good communication between all members of the building team. As it will fall to the quantity surveyor to maintain construction cost records and provide the employer with regular financial reviews, it is essential that he attends all meetings when cost matters are to be discussed and that he receives copies of all documents that may have some bearing on the cost of the project.

It is also essential for the cost effect of all proposed variations, whether emanating from the architect or from any of the other design consultants, to be determined so that any necessary corrective action can be taken to minimise their impact on the contract sum and to maintain overall expenditure within authorised limits before the architect's instructions authorising them are issued. Traditionally this has been done, with varying degrees of accuracy, by the quantity surveyor estimating the cost effect of proposed variations. However, with the incorporation into JCT98 of the contractor's price statement and the contractor's 13A quotation, it is now possible to obtain a fixed-price quotation from the contractor for a proposed variation prior to authorising its execution. Whilst using these procedures may take a little longer than the quantity surveyor's estimate, they do provide fixed costs, which must lead to more accurate cost control.

By looking ahead and making early decisions on such matters as the nomination of sub-contractors and suppliers and the expenditure of provisional sums, the architect can greatly assist the cost control process.

The regular financial reviews referred to above will normally be produced by the quantity surveyor and sent to the employer and other members of the design team at monthly intervals, often coinciding with the dates of issue of interim certificates. These reviews should show a forecast of the employer's total financial commitment to the contractor, including the reimbursement of direct loss and/or expense but normally excluding VAT, and should, in effect, comprise an estimate of the final adjustment of the contract sum. A list of all the adjustments to be made to the contract sum is set down in JCT98 clause 30.6.2 and this can act as a most useful *aide memoire* when preparing a financial review.

A typical financial review is given as Example 13/2 at the end of this chapter.

Issued by: Ivor Barch Associates address: Prospects Drive, Fairbridge	**Architect's Instruction**

Employer: Cosmeston Preparatory School address: Fairbridge	Job reference: IBA/94/20
	Instruction no: 4
Contractor: L&M Construction Ltd address: Ferry Road, Fairbridge	Issue date: 18 February 1999
	Sheet: 1 of 2
Works: New School Library situated at: Park Street, Fairbridge	

Contract dated: 14 December 1998

Under the terms of the above-mentioned Contract, I/we issue the following instructions:

	Office use: Approximate costs	
	£ omit	£ add
1. **INCOMING GAS MAIN** Accept the quotation ref. no. 8438/63 dated 6 January 1999 received from EuroGas in the sum of £248.00 for the new incoming gas main and supply and installation of gas meter. A copy of their quotation is attached. This sum is to be set against the provisional sum of £400 included under ref. 3/2G in the bill of quantities.	400.00	248.00
2. (a) **HIP TILES** OMIT: Farland concrete third round hip tiles, bill of quantities ref. 5/15E. ADD in lieu: Red Bank 300mm long red Terracotta third round segmental ridge tiles, list no. 259. (b) **HIP IRONS** OMIT: 4 no. hip irons, bill of quantities ref. 5/15F ADD in lieu: 4 no. Red Bank 300mm long red Terracotta Scroll hip finial tiles 225mm diameter. Hip tiles and finial tiles to be obtained from Farbridge Bank Manufacturing Co. Ltd.	372.00 –	514.00 –
3. **PICTURE-HANGING FACILITIES IN ACTIVITIES ROOM** Supply and fix aluminium picture rail approx. 12m long and 100 no. type (b) U-shaped hooks obtainable from Library Aids of Thawbridge. Picture rail to be fixed in location shown on attached drawing no. IBA/94/20/61 at 2050mm from finished floor level. <div align="center">(continued)</div>	–	80.00

To be signed by or for the issuer named above	Signed *Ivor Barch*	sub-total:	772.00	842.00

Amount of Contract Sum	£	
± Approximate value of previous Instructions	£	
Sub-total	£	
± Approximate value of this Instruction	£	
Approximate adjusted total	£	

Distribution	☐ Contractor	☐ Quantity Surveyor	☐ Clerk of Works	☐
	☐ Employer	☐ Structural Engineer	☐ Planning Supervisor	☐
	☐ Nominated Sub-Contractors	☐ M&E Consultant	☐	☐ File

F809   for JCT 98 / IFC 98 / MW 98   © RIBA Publications 1999

Example 13/1   Architect's instruction. (*Continued.*)

		£ omit	£ add
	Brought forward:	772.00	842.30
4.	VELUX ROOFLIGHTS OMIT: single glazing with 6mm thick clear polycarbonate sheet, bill of quantities ref. 5/25B. ADD in lieu: single glazed in 10mm thick Clerestory Glass Ltd. 'Kevlarplate' float clear.	–	–
5.	MORTAR MIXES This is to confirm Clerk of Works Direction no. 6 that mortar mixes are as follows: Blockwork below dpc 1:3 Blockwork above dpc 1:1:6 Brickwork below dpc (below ground level) 1:3 Brickwork below dpc (above ground level) 1:3 (pointing in 2:1:9) Brickwork above dpc 2:1:9 Coloured mortar is to be Zilcon mortar, ref. Y73.	–	–
6.	REDUCED LEVELS This is to confirm Clerk of Works Direction no. 6. Excavate to revised levels to remove unsuitable material. The 'as dug' levels as shown on the record made by the site foreman and clerk of works are agreed.	–	–

Issued by: Ivor Barch Associates
address: Prospects Drive, Fairbridge

Instruction
continuation

Job reference: IBA/94/20

Instruction no: 4

Issue date: 18 February 1999

Sheet: 2   of 2

This instruction is issued in accordance with clauses 13.3, 13.2 and 4.1 of the contract.

(Items nos. 4, 5 and 6 are to be at no cost to the Employer.)

To be signed by or for the issuer named above	Signed *Ivor Barch*	772.00	842.30

Amount of Contract Sum	£	505,096.00
± Approximate value of previous Instructions	£	500.00
Sub-total	£	505,596.00
± Approximate value of this Instruction	£	70.00
Approximate adjusted total	£	505,666.00

F820   for JCT 98 / IFC 98 / MW 98

© RIBA Publications 1999

Example 13/1   *(Continued.)*

FUSSEDON KNOWLES & PARTNERS
Chartered Quantity Surveyors
Upper Market Street
Borchester BC2 1HH

Financial Review No: 9
Date of Issue: 12 March 2002
Reference: 1234

Works: Shops & Offices, Newbridge Street, Borchester

Contractor: Leavesden Barnes & Co Ltd

Employer: Aqua Products Ltd

Contract Sum: £675,332.00

Approved Expenditure: £680,000

Date of Possession: 11 June 2001

Date for Completion: 26 July 2002

ESTIMATED CURRENT FINANCIAL COMMITMENT MORE THAN APPROVED EXPENDITURE

	£	£
Contract sum	675,332	
Deduct contingencies	20,000	
		655,332
Add estimated value of variations		23,980
		679,312
Add/deduct estimated fluctuations in cost of labour and materials		NIL
		679,312
Add estimated reimbursement of direct loss and/or expense		2,000
Total estimated final cost		681,312
Deduct approved expenditure		680,000
OVERSPENT BY	£	**1,312**

ESTIMATED CURRENT FINANCIAL COMMITMENT LESS THAN APPROVED EXPENDITURE

	£	£
Approved expenditure		
Contract sum		
Deduct contingencies		
Add/deduct estimated value of variations		
Add/deduct estimated fluctuations in cost of labour and materials		
Add estimated reimbursement of direct loss and/or expense		
Total estimated final cost		
UNDERSPENT BY	£	

Notes: 1. All amounts given above are EXCLUSIVE of fees and Value Added Tax

---

Signature:

Date: 12 March 2002

Example 13/2   Financial review.

Chapter 14
Interim Payments

Introduction

Prior to 1998 it was a basic principle of English contract law that when a contractor undertook to do work for a fixed sum, he was not due any payment until the whole of the work had been completed, but once the work was complete, he was due full payment of the fixed sum. Whilst this arrangement may have been appropriate for small projects, construction work can involve large sums of money being expended over a number of months or even years. Thus, in order to provide cash flow to the contractor, this basic principle was usually varied by agreement between the parties to allow interim payments to be made as the work proceeded.

The Housing Grants, Construction and Regeneration Act 1996 (Construction Act) and The Scheme for Construction Contracts (England and Wales) Regulations 1998 (The Scheme) came into effect on 1 May 1998 and changed contract law in respect of what the Construction Act defined as construction contracts, except for those with a residential occupier. Amongst other things, the Construction Act requires that in respect of payments:

- A party to a construction contract is entitled to payment by instalments, stage payments or other periodic payments for any work under the contract unless the duration of the work is less than 45 days. The parties are free to agree the amounts of the payments and the intervals at which, or the circumstances in which, they become due. In the absence of agreement such amounts, intervals or circumstances have to be determined by reference to Part II of The Scheme.
- Every construction contract shall provide a mechanism for determining what payments become due under the contract, and when, and shall also provide for a final date for payment of any sum which becomes due. The parties are free to agree how long the period is to be between the date on which the sum becomes due and the final date for payment. In the absence of such provisions they have to be determined by reference to Part II of The Scheme.

- Every construction contract shall provide for the giving of notice by a party, not later than five days after the date on which a payment becomes due from him, specifying the amount of the payment to be made and the basis on which that amount is calculated. In the absence of such provisions, notice specifying the amount of payment has to be that set down in Part II of The Scheme.
- A party to a construction contract may not withhold payment after the final date for payment unless he has given an effective notice of intention to withhold payment not later than the prescribed period before the final date for payment. The parties are free to agree what the prescribed period is to be. In the absence of such agreement, the period has to be that set down in Part II of The Scheme.
- Where the sum due under a construction contract is not paid in full by the final date for payment and no effective notice to withhold payment has been given, the person to whom the sum is due has the right to suspend performance of his obligations under the contract.
- A provision making payment under a construction contract conditional on the payer receiving payment from a third party is ineffective, unless that third party is insolvent.

Most standard forms of contract now incorporate such payment provisions as are necessary to comply with the Construction Act. They also provide for money to be held in trust, as retention both during and at the end of the period of construction, primarily to provide a fund from which the employer is able to recover the cost of having defects repaired which the contractor may be unwilling to put right.

When considering the subject of interim payments it is worth bearing in mind that the contractor has similar obligations to his suppliers and sub-contractors as does the employer to the contractor. Provided that the minimum requirements of the Construction Act are met, interim payments can take one of the following forms:

- payments of pre-determined amounts at regular intervals
- payments of pre-determined amounts when the work reaches pre-determined stages of completion
- payments of amounts at regular intervals calculated by a detailed valuation process as the work proceeds.

Payments of pre-determined amounts at regular intervals

This is a method where, for example, a construction project of £120,000 is contracted to take 12 months and it is determined that the contractor will be paid £10,000 per month, adjusted to account for authorised variations, fluctuations, loss and/or expense and retention. The method has the advantage, to both parties, of minimal administration and cash flow certainty. However, it can lead to problems if the contractor falls behind programme or executes defective work, whereupon over-payment to the contractor for the work done is a probability, which would certainly lead to problems in the event of the contractor going into liquidation before completing the project.

Payments of pre-determined amounts when the work reaches pre-determined stages of completion

This is a method whereby pre-determined amounts, adjusted to account for authorised variations, fluctuations, loss and/or expense and retention, become due for payment upon the proper completion of a pre-determined stage of work. For example, the pre-determined amount of, say, £26,000 (adjusted as noted above) becomes payable upon the proper completion of all work up to damp-proof course level.

Determining the amounts of interim payments in this way keeps down the parties' administration costs and, unlike the previous method, does not lead to overpayment to the contractor if he falls behind programme or if he executes defective work. It can, however, be disadvantageous to the contractor as it does not take into account the value of unfixed materials and goods either on or off site. Also the non-completion of a minor, non-critical and low-value element of work in a high-value stage would prevent payment becoming due for the whole stage.

It is interesting to note that JCT98 With Contractor's Design provides for the optional use of this method of payment under clause 30.2A Alternative A – Stage Payments.

Payments of amounts at regular intervals calculated by a detailed valuation process as the work proceeds

This method of determining the amounts of interim payments is the fairest and most commonly adopted in construction contracts. The implications of defective work, delays, authorised variations, materials on and off site, fluctuations, loss and expense and retention can all be readily taken into account in each payment. This procedure does, however, require a regular and substantial input from most members of the building team.

JCT98 adopts this method of payment and sets out the procedures to be followed in considerable detail in clause 30 – Certificates and Payments.

Certificates and payments under JCT98

The obligations of the employer's consultants and the parties to the contract, as set out in clause 30, are briefly as follows.

The architect

The architect has to issue interim certificates stating the amount due to the contractor, to what that amount relates and the basis of calculation of that amount:

- on the dates provided for in the appendix to the contract up to the date of practical completion or to within one month thereafter
- as and when further amounts are ascertained as being payable to the contractor after practical completion
- after expiration of the defects liability period or upon issue of the certificate of completion of making good defects, whichever is the later
- as soon as practicable, and in any event not less than 28 days before the issue of the final certificate – the gross valuation of this interim certificate has also to include the ascertained amounts of the final accounts of all nominated sub-contractors.

Except in the case of the interim certificate including the ascertained amounts of the final accounts of all nominated sub-contractors, the architect is not required to issue interim certificates at intervals of less than one calendar month. When issuing his interim certificates, the architect must direct the contractor as to the amounts included for nominated sub-contractors and he must notify each nominated sub-contractor of the amount included for them.

Interim certificates are vital to the smooth running of a project and the architect should bear in mind that:

- he must issue his certificate at the appointed time and in accordance with the requirements of the contract
- he is legally responsible for the accuracy of his certificate
- he must remain independent in the issuing of certificates so as to be fair to both parties.

The quantity surveyor

If the contractor submits an application setting out his gross valuation to the quantity surveyor, the quantity surveyor has to make an interim valuation. If the quantity surveyor disagrees with the gross valuation of the contractor

... the architect is legally responsible for the accuracy of his certificates
...

(and/or nominated sub-contractors), he has to submit a statement to the contractor at the time of making his valuation and in similar detail to the contractor's application, identifying his disagreement. If clause 40 – use of price adjustment formulae – applies, the quantity surveyor has to make an interim valuation before the issue of each interim certificate. Other than on these two occasions it is at the architect's discretion whether or not interim valuations are to be made by the quantity surveyor.

The contractor

Whilst clause 30.2 allows the contractor to submit to the quantity surveyor an application setting out his gross valuation pursuant to clause 30.2 (and any nominated sub-contractors' gross valuations pursuant to clause 4.17 of Conditions NSC/C), he is under no contractual obligation to assist, in any way, in the preparation of interim valuations or certificates. It is the architect's responsibility to issue the interim certificates at the correct times and it is the employer's responsibility to make payment to the contractor within the 14 days period.

Not later than five days after the date of issue of an interim certificate the contractor ought to give written notice to each nominated sub-contractor specifying the amount of the payment in respect of their sub-contract works included in the amount certified, to what that amount relates, and the basis of calculation of that amount. Not later than five days before the final date for payment of the amount certified due, the contractor may give written notice to each nominated sub-contractor specifying the amount proposed to be withheld or deducted from the amount due to them, the ground(s) for so doing, and the amount attributable to each ground. If the contractor does not give these notices and fails to pay, in full, the amount stated as due in an interim certificate by the final date for payment, that is within 17 days from the date of issue of the interim certificate, then:

- the nominated sub-contractor is entitled to be paid simple interest on the overdue amount at 5% over base rate of the Bank of England
- the nominated sub-contractor is empowered, subject to giving the notice required by clause 4.21.1 of NSC/C, to suspend the performance of his obligations under the contract until payment in full occurs
- the employer shall, subject to the issue by the architect of the certificate required by and to the provisos contained in clause 35.13, pay the nominated sub-contractor direct (see 'Payments to nominated sub-contractors under JCT98' later in this chapter).

The employer

Not later than five days after the date of issue of an interim certificate the employer ought to give written notice to the contractor (in respect of the amount certified due) specifying the amount of the payment proposed, to what that amount relates, and the basis of calculation of that amount. Not later than five days before the final date for payment of the amount certified due, the employer may give written notice to the contractor specifying the amount proposed to be withheld or deducted from the due amount, the ground(s) for so doing, and the amount attributable to each ground.

If the employer does not give these notices and fails to pay, in full, the amount stated as due in an interim certificate by the final date for payment, that is within 14 days from the date of issue of the interim certificate, then:

- the contractor is entitled to be paid simple interest on the overdue amount at 5% over base rate of the Bank of England

- the contractor is empowered, subject to giving the notice required by clause 30.1.4, to suspend the performance of his obligations under the contract until payment in full occurs
- the contractor is empowered, subject to giving the notices required by clause 28.2, to determine his employment under the contract. He may also start proceedings in the courts for the recovery of the debt.

Whilst the employer does have the right, in the circumstances prescribed by and subject to giving the notices required by clause 24, to deduct liquidated and ascertained damages from the amount stated as due in an interim certificate, he does not have the right to interfere with the issue of any architect's certificate. If he does so it is a matter for which the contractor may determine his employment under the contract. However, the contractor would achieve nothing by determining if it were the final certificate that was obstructed or interfered with, and it is doubtful whether he would achieve much by determining close to or after practical completion. In such circumstances he would, instead, be better off invoking the procedures for the settlement of disputes or differences prescribed by clauses 41A and 41B or 41C.

When it is recorded in the appendix to the contract that the employer is a 'contractor' for the purposes of the Construction Industry Scheme, the employer must comply with the provisions of clause 31 when making payment to the contractor, including making statutory deductions from that part of the payment which is not in respect of the direct cost of materials and accounting to the Inland Revenue for those deductions.

Interim certificates under JCT98

Whether or not there is a contractual requirement for the quantity surveyor to prepare interim valuations for the purpose of ascertaining the amount to be stated as due in an interim certificate, it is normal practice on most projects for him to do so and for the contractor to co-operate with him in performing that task.

According to clause 30.2 the amount to be stated as due in an interim certificate is the gross valuation up to and including a date not more than seven days before the date of the interim certificate less:

- any amount which may be deducted by the employer as the retention
- the total amount of any advance payment due for reimbursement to the employer
- the total amount stated as due in previous interim certificates.

Clause 30.2 also provides that the gross valuation shall be:

- the total of all the amounts to be included which are subject to retention
- plus the total of all the amounts to be included which are not subject to retention
- less the total of all the amounts to be deducted which are not subject to retention.

The amounts to be included which are subject to retention are:

- the total value of the work, *properly executed* by the contractor, including variations and, where applicable, any adjustment for fluctuations by use of the price adjustment formulae
- the total value of materials and goods delivered to, or adjacent to the works and for incorporation therein by the contractor
- the total value of materials or goods or items pre-fabricated, which have been listed by the employer, before their delivery to site
- the total value of the above three points but in respect of each nominated sub-contractor
- the profit of the contractor upon the total amount included for each nominated sub-contractor, including their fluctuations and any other monies due to them under the terms of their nominated sub-contract.

The amounts to be included which are not subject to retention are:

- any amount which becomes due to the contractor under the terms of the contract in respect of statutory fees and charges, opening up work for inspection and testing, royalties arising out of compliance with an architect's instruction, insurances under clauses 21.2.3 and 22D.4 and costs incurred if the employer defaults under clauses 22B or 22C
- any amount due to the contractor by way of reimbursement of loss and/or expense arising from matters materially affecting the regular progress of the works or from the discovery of antiquities
- any amount due to the contractor for the replacement or repair of loss or damage and removal and disposal of debris which is treated as a variation
- the amount of any final payment to a nominated sub-contractor
- any amount due to the contractor for fluctuations calculated other than by use of the price adjustment formulae
- any amount properly payable to a nominated sub-contractor for the replacement or repair of loss or damage and removal and disposal of debris which is treated as a variation and for fluctuations calculated other than by use of the price adjustment formulae.

The amounts to be deducted which are not subject to retention are:

- any amount in respect of errors in setting out which the architect instructs are not to be amended
- any amount in respect of work not in accordance with the contract which the architect may allow to remain
- any amount in respect of any defects, shrinkages or other faults which the architect instructs are not to be made good
- any amount allowable by the contractor or a nominated sub-contractor for fluctuations calculated other than by use of the price adjustment formulae.

Unfixed materials and goods on site

The value of all materials and goods stored on site must be included in the valuation, provided that they are adequately protected and have not been brought to site prematurely. For example, unless there was some prior agreement on the matter, the architect is entitled to withhold payment for items of furniture brought onto site whilst only work in the foundations is in progress. Any materials or goods included in the amount stated as being due in an interim certificate that has been paid by the employer become the property of the employer and, although the contractor remains responsible for their loss or damage, he must not remove them from the site.

Unfixed materials and goods off site

Only the value of materials or goods stored off site that have been listed by the employer in a list supplied to the contractor and annexed to the contract documents can be considered for inclusion in the gross valuation, and then only if the following criteria are met:

- the contractor has provided the architect with reasonable proof that the property in the listed items is vested in the contractor
- if so stated in the appendix to the contract, the contractor has provided a bond in favour of the employer from a surety approved by the employer in respect of payment for the listed items
- the listed items are in accordance with the contract
- the listed items are set apart or are clearly and visibly marked and identify the employer and to whose order they are held and their destination as the works
- the contractor has provided the employer with reasonable proof that the listed items are insured against loss or damage for their full value.

Any listed items included in the amount stated as being due in an interim certificate that has been paid by the employer become the property of the employer and, whilst the contractor remains responsible for their loss or damage, and the cost of their storage, handling and insurance, he must not remove them from the premises where they are stored except for the purpose of their delivery to site.

Retention under JCT98

The contract provides that, in the calculation of the amount to be stated as due in an interim certificate, the employer may deduct and retain the retention from the gross valuation. The total amount of the retention that may be deducted and retained at any one time is ascertained by application of the rules given in clause 30.A and comprises:

- the total value of work, including that of nominated sub-contractors, which has not reached practical completion and of materials and goods included in the gross valuation – at the retention percentage
- the total value of work, including that of nominated sub-contractors, which has reached practical completion or has been taken into the employer's possession by agreement with the contractor but for which a certificate of making good defects has not been issued, included in the gross valuation – at half the retention percentage.

The retention percentage is 5% unless a lower rate is specified in the appendix to the contract. It is to be noted that no retention is held against the value of work for which a certificate of making good defects has been issued.

The treatment of the retention is subject to the rules given in clause 30.5 which provide that:

- the employer's interest in the retention is fiduciary as trustee for the contractor and any nominated sub-contractor, but without an obligation to invest. In other words it is money held by the employer in trust for the contractor and any nominated sub-contractor
- the architect or the quantity surveyor, if so instructed by the architect, must prepare a statement of the contractor's retention and of the nominated sub-contract retention for each nominated sub-contractor at the date of each interim certificate – the architect has to issue copies of that statement to the employer, the contractor and each nominated sub-contractor
- the employer shall, at the request of the contractor or any nominated sub-contractor, place the retention in a separate, appropriately designated banking account and certify to the architect, with a copy to the contractor, that this has been done (this rule does not appear in the Local Authorities edition of JCT98)

- when the employer exercises his right to withhold and/or deduct monies due to him under the terms of the contract against any retention, he must inform the contractor of the amount withheld and/or deducted from either the contractor's retention or the nominated sub-contract retention of any nominated sub-contractor.

Payments to nominated sub-contractors under JCT98

In the normal course of events Clause 35.17 imposes an obligation on the employer to make early final payment to nominated sub-contractors. This obligation arises under the terms of Agreement NSC/W, which is one of the standard documents to be used when nominating sub-contractors (see Chapter 8).

Clause 35.16 requires that when, in the opinion of the architect, practical completion of the works executed by a nominated sub-contractor is achieved, he shall immediately issue a certificate to that effect. So long as Agreement NSC/W clause 5 remains in force and is unamended, clause 35.17 requires that at any time after the date of practical completion of the nominated sub-contract works, the architect may – and on the expiry of 12 months from the date of practical completion of that work, the architect must – issue an interim certificate, the gross valuation for which must include the finally adjusted or ascertained nominated sub-contract sum. The issue of this interim certificate is subject to the nominated sub-contractor:

- having, in the opinion of both the architect and the contractor, remedied any defects, shrinkages or other faults
- having sent through the contractor to the architect or the quantity surveyor all documents necessary for the final adjustment of the nominated sub-contract sum.

Before the issue of each interim certificate, other than the first, the contractor must provide the architect with reasonable proof that he has made payments to nominated sub-contractors due under previous interim certificates. If he is unable to provide such proof, the procedures for direct payment by the employer will come into force unless the contractor is able to satisfy the architect that the absence of proof is due to some failure or omission by the nominated sub-contractor concerned. The procedure for direct payment of the nominated sub-contractor is mandatory on the employer if Agreement NSC/W has been entered into.

If reasonable proof of payment is not forthcoming the architect must issue a certificate to that effect, stating the amount in respect of which the contractor has failed to provide such proof. He must send a copy of that certificate to the nominated sub-contractor concerned. Provided that the certificate has been issued, the amount of any future payment otherwise due to the contractor will be reduced by the amount by which the contractor has defaulted in his payment to the nominated sub-contractor and the employer will pay the nominated sub-contractor concerned direct.

Any direct payment is to be made when the employer makes payment to the contractor in respect of the next interim certificate. If, after deduction of the amount of the direct payment from the amount stated as being due in the interim certificate, there is no balance due to the contractor, the direct payment to the sub-contractor must be made during the 14-day period within which the contractor would have been paid.

There is no obligation on the employer to make direct payments to nominated sub-contractors in excess of the amount certified due to the contractor. If the amount due to the contractor is all or a part of the retention, the amount of the direct payment and the amount by which the payment to the contractor is reduced must not exceed any part of the contractor's retention which would otherwise be due for payment to the contractor.

If there is more than one nominated sub-contractor to be paid direct, and the amount due to the contractor is insufficient to meet the total of the direct payments, the employer must apportion the amount available pro rata to the undischarged amounts, or on some other fair and reasonable basis. If the contractor goes into liquidation the provisions for direct payments to nominated sub-contractors immediately cease to have effect.

Before making a direct payment to any nominated sub-contractor the employer must remember to comply with the requirements of clause 30.1.1.4 with regard to giving the contractor notice of the proposed deduction from the amount certified as being due.

Value added tax

Clause 15.2 makes it quite clear that the contract sum is exclusive of value added tax. The responsibility for the payment of this tax lies with the contractor and he in turn will invoice the employer appropriately. Clause 15.2 gives the contractor the right to recover VAT from the employer and the provisions for that recovery are set out in the supplement to the conditions of contract, known as the VAT Agreement. The VAT Agreement sets out two alternatives for the contractor to recover VAT. The appendix to the conditions of contract requires Clause 1A of the VAT Agreement either to apply or not to apply. If clause 1A is to apply, the architect must show the VAT properly due on each certificate.

Valuation and certificate forms

Except where employers have their own forms on which they require interim payments to be certified, it is normal practice for standard forms published by the professional bodies to be used. Valuation forms are published by the RICS for the use of quantity surveyors. These comprise the valuation shown in Example 14/1 and the statement of retention and of nominated sub-contractors' values, which accompanies the valuation form as an appendix, shown in Example 14/2.

RIBA Publications publishes forms for architects to use in connection with interim certificates. These are the interim certificate, the statement of retention and of nominated sub-contractors' values and the notification to nominated sub-contractors of amounts included for them in the certificate. The first two of these forms can be filled in directly from the information given in the quantity surveyor's valuation – completed specimens are shown in Examples 14/3 and 14/4. The notification form is shown in Example 14/5. All these forms are published in separate pads and each pad contains notes on their use. It will also be seen that the quantity surveyor's valuation form contains several notes which form a useful reminder of the contractual obligations.

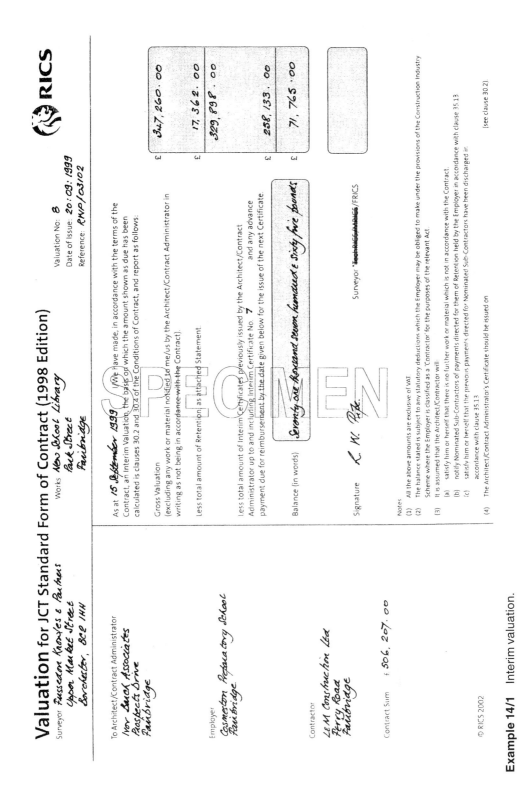

Valuation for JCT Standard Form of Contract (1998 Edition)

Surveyor *Fussston Knowles & Partners*
Upon Market Street
Dorchester, BC2 1HH

Works *New School Library*
Park Street
Fairbridge

Valuation No: **8**
Date of Issue: **20 : 09 : 1999**
Reference: **RNP/03/02**

To Architect/Contract Administrator
Ivor Burol Associates
Prospects Drive
Fairbridge

Employer
Cosmerton Preparatory School
Fairbridge

Contractor
LEM Construction Ltd
Perry Road
Fairbridge

Contract Sum £ *506, 207. 00*

As at *15 September 1999* I/We have made, in accordance with the terms of the Contract, an Interim Valuation, the basis of which the amount shown as due has been calculated is clauses 30.2 and 30.4 of the Conditions of Contract, and report as follows:

Gross Valuation
(excluding any work or material notified to me/us by the Architect/Contract Administrator in writing as not being in accordance with the Contract).   £ **347,260 . 00**

Less total amount of Retention as attached Statement.   £ **17,362 . 00**

                                                                                                £ **329,898 . 00**

Less total amount of Interim Certificates previously issued by the Architect/Contract Administrator up to and including Interim Certificate No. **7**   and any advance payment due for reimbursement by the date given below for the issue of the next Certificate.   £ **258,133 . 00**

Balance (in words) *Seventy one thousand seven hundred & sixty five pounds*   £ **71, 765 . 00**

Signature *K. W. Pye.*   Surveyor ~~Technical Manager~~/FRICS

Notes:
(1)   All the above amounts are exclusive of VAT
(2)   The balance stated is subject to any statutory deductions which the Employer may be obliged to make under the provisions of the Construction Industry Scheme where the Employer is classified as a 'Contractor' for the purposes of the relevant Act.
(3)   It is assumed that the Architect/Contractor will:
        (a)   satisfy him or herself that there is no further work or material which is not in accordance with the Contract.
        (b)   notify Nominated Sub-Contractors of payments directed for them of Retention held by the Employer in accordance with clause 35.13
        (c)   satisfy him or herself that the previous payments directed for Nominated Sub-Contractors have been discharged in accordance with clause 35.13
(4)   The Architect/Contract Administrator's Certificate should be issued on   (see clause 30.2).

© RICS 2002

Example 14/1   Interim valuation.

Statement of Retention and of Nominated Sub-Contractors' Values

Surveyor: *Busedon Knowles & Partners, Upper Market Street, Borchester, BC2 1HH*

Works: *New School Library, Park Street, Fairbridge*

This Statement relates to:
Valuation No: *8*
Date of Issue: *20·09·1999*
Reference: *RWP/03/02*

RICS

	Gross Valuation £	Basis of Gross Valuation (See note 1) Clause No.	Amount Subject to. Full Retention of % £	Half Retention of % £	No Retention £	Amount of Retention £	Net Valuation £	Amount Previously Certified £	Balance £
Main Contractor	334,210.00		334,210.00	–	–	16,710.00	317,500.00	258,133.00	59,367.00
Nominated Sub-Contractors									
1. *Speediweld (UK) Ltd* (steelwork)	2,700.00	4.17	2,700.00	–	–	135.00	2,565.00	–	2,565.00
2. *Nor-Short Ltd* (electrical)	1,450.00	4.17	1,450.00	–	–	72.00	1,378.00	–	1,378.00
3. *Pinkline Ltd* (architectural metalwork)	8,900.00	4.17	8,900.00	–	–	445.00	8,455.00	–	8,455.00
TOTAL	347,260.00		347,260.00	–	–	17,362.00	329,898.00	258,133.00	71,765.00

Notes:
(1) The basis of the gross valuation in respect of Nominated Sub-Contractors is
     for interim payments — clause 4.17 of NSC/C   — (clause 3.1 of NSC/A) clause 4.23 of NSC/C
     for final payments — sub-contracts based on a lump sum — (clause 3.1 of NSC/C)
                          — sub-contracts based on measurement — (clause 3.2 of NSC/A) clause 4.24 of NSC/C
(2) No account has been taken of any discounts for cash which the Contractor may be entitled if discharging the balance within 17 days of the issue of the
     Architect/Contract Administrator
(3) The sums stated are exclusive of VAT.

© RICS 2000

Example 14/2　　Statement of retention.

**Interim
Certificate**

and Direction

JCT 98

Issued by: Ivor Barch Associates
address: Prospects Drive, Fairbridge

Employer: Cosmeston Preparatory School   Serial no: **G**
address: Fairbridge

Job reference: IBA/94/20

Contractor: L&M Construction Ltd   Certificate no: 8
address: Ferry Road, Fairbridge

Date of valuation: 15 September 1999

Works: New School Library   Date of issue: 23 September 1999
situated at: Park Street, Fairbridge

Final date for payment: 7 October 1999

Contract dated: 14 December 1998

Original to Employer

This Interim Certificate is issued under the terms of the above-mentioned Contract.

Gross valuation .... £ 347,260.00

Less Retention as detailed on the Statement of Retention .... £   17,363.00

Sub-total £ 329,897.00

Less reimbursement of advance payment .... £   –

Sub-total £ 329,897.00

Less total amount previously certified .... £ 258,133.00

Net amount for payment .... £   71,764.00

I/We hereby certify that the **amount due** to the Contractor from the Employer is (in words)   *All amounts are exclusive of VAT.*

Seventy One Thousand Seven Hundred and Sixty Four Pounds

I/We hereby direct the Contractor that this amount includes interim or final payments to Nominated Sub-Contractors as listed in the attached *Statement of Retention and of Nominated Sub-Contractor's Values,* which are to be paid to those named in accordance with the Sub-Contract.

To be signed by or for
the issuer named
above

Signed *Ivor Barch*

[1] Relevant only if
clause 1A of the VAT
Agreement applies.
Delete if not
applicable.

[1] The Contractor has given notice that the rate of VAT chargeable on the supply of goods and services to which the Contract relates is   17.5 %

[1]   17.5 % of the amount certified above .... £ 12,558.70

[1] Total of net amount and VAT amount (for information) .... £ 84,322.70

This is not a Tax Invoice.

F801   for JCT 98   © RIBA Publications 1999

Example 14/3   Interim certificate.

Issued by: Ivor Barch Associates
address: Prospects Drive, Fairbridge
Works: New School Library
situated at: Park Street, Fairbridge

Statement of Retention
and of Nominated Sub-Contractor's Values

Job reference: IBA/94/20
Relating to Certificate no: 8
Issue date: 23 September 1999

	Gross valuation £	Amount subject to: Full retention of % £	Half retention of % £	Nil retention £	Amount of retention £	Net valuation £	Previously certified £	Balance due £
Main Contractor	334,210.00	334,210.00		–	16,710.00	317,500.00	258,133.00	59,367.00
Nominated Sub-Contractors:								
1 Speediweld (UK) Ltd (steelwork)	2,700.00	2,700.00		–	135.00	2,565.00	–	2,565.00
2 Ivor Short Ltd (electrical)	1,450.00	1,450.00		–	72.00	1,378.00	–	1,378.00
3 Fineline Ltd (architectural metalwork)	8,900.00	8,900.00		–	445.00	8,455.00	–	8,455.00
Total	347,260.00	347,260.00		–	17,362.00	329,898.00	258,133.00	71,765.00

All amounts are exclusive of VAT.
F802 for JCT 98

We are grateful to the Royal Institution of Chartered Surveyors for allowing us to adapt their copyright form of the same title.
© RIBA Publications 1999

Example 14/4   Statement of retention.

Notification
to Nominated
Sub-Contractor
of amount included
in certificate

Issued by: Ivor Barch Associates
address: Prospects Drive, Fairbridge

Employer: Cosmeston Preparatory School
address: Fairbridge

Job reference: IBA/94/20

Notification no: 1

Main Contractor: L&M Construction Ltd
address: Ferry Road, Fairbridge

Issue date: 23 September 1999

Works: New School Library
situated at: Park Street, Fairbridge

Contract dated: 14 December 1998

Original to Nominated
Sub-Contractor

Nominated
Sub-Contractor: Speediweld (UK) Ltd. Fairbridge
address:

Under the terms of the above-mentioned Main Contract,

I/we hereby certify that I/we have directed the Contractor that

Interim Certificate no. 8 dated 23 September 1999

*Delete as
appropriate

includes *an interim/a final payment of .... £ 2,565.00
which is to be paid to you.

SPECIMEN

To be signed by or for
the issuer named
above

Signed _____ *Ivor Barch* _____

- -

Nominated Sub-Contractor's
Proof of Payment
of amount due

Main Contractor:
address:

Works:
situated at:

Job reference:

In accordance with the terms of the relevant Sub-Contract,
we confirm that we have received from you payment of the amount of ...... £ _____

included in Interim Certificate no. _____ dated _____

as stated in Notification no. _____ dated _____

Please complete
Proof of Payment
slip and send to
Contractor

Signed _____ Date _____

for _____

F803 for JCT 98

© RIBA Publications 1999

Example 14/5 Notification to Nominated Sub-Contractor.

Chapter 15

Completion, Defects and the Final Account

A building contract does not come to an end until the architect issues his final certificate and even then actions for breach of contract can be commenced within 6 or 12 years of the breach, depending upon whether the contract had been executed under hand or as a deed (see Chapter 10). Completion of the works can take place in three stages:

- practical completion
- sectional completion (if the JCT Sectional Completion Supplement is applicable) or partial possession
- completion of making good defects.

Practical completion

In JCT98 practical completion of the works is defined by reference to clause 17.1. When in the opinion of the architect the contractor:

- has achieved practical completion of the works, including all authorised variations
- has provided such information as is reasonably required by the planning supervisor for the preparation of the health and safety file required by CDM94
- has supplied the employer with the required drawings and information showing and/or describing any performance-specified work as built and, where relevant, its maintenance and operation

then the architect is required to issue a certificate to that effect and practical completion of the works will, for the purposes of the contract, be deemed to have taken place on the day named in that certificate. This date is significant as it has an important influence on a number of conditions in the contract and disputes that may arise under it. RIBA Publications publishes a standard pro forma Certificate of Practical Completion; a completed copy of this form is given as Example 15/1 at the end of this chapter.

Over the years the courts have defined practical completion of the works in differing ways but the safest approach for any architect to adopt is to certify practical completion of the works only when every item of work has been satisfactorily completed. However, the architect is often put under pressure to certify practical completion prematurely to enable the employer to gain occupation of his building. The consequences to the employer and the contractor of issuing the certificate of practical completion whilst items of work remain incomplete, even if they are of a minor nature, must be considered by the architect – for example, the contractor will have to complete work in an occupied (and perhaps otherwise finished) building, giving him the additional problems related to health and safety and protection. If an architect were minded to issue his certificate of practical completion when items of work remain incomplete, he would be well advised to separately identify, in writing to the contractor, the items of work that remain to be completed, an order of priority for their completion and the date(s) by when they are to be completed. Practical completion ought *never* to be certified when there is any outstanding *defective* work (as distinct from incomplete work).

The achievement of practical completion is used in JCT98 as the datum for the effective commencement or cessation of various provisions. These are:

Clause	Description
1.5	The contractor remains wholly responsible for carrying out and completing the works in accordance with the contract whether or not the architect issues the certificate of practical completion.
5.9	The contractor has to supply the employer with the required drawings and information showing and/or describing any performance specified work as built and, where relevant, its maintenance and operation, before the date of practical completion.
17.1	The architect must issue a certificate of practical completion of the works.
17.2	The defects liability period (normally 6 months) runs from the day named in the certificate of practical completion of the works.
17.3 & 17.5	Defects, shrinkages or other faults caused by frost occurring before practical completion of the works, but not thereafter, have to be made good by the contractor at no cost to the employer.
19.1.2	Where the appendix states that this clause applies the employer may, at any time after practical completion of the works, assign to a transferee or lessee the right to bring proceedings in the name of the employer to enforce any of the terms of the contract made for the benefit of the employer.
21.2.1.5	Except for any part(s) of the works that are the subject of a certificate of practical completion, injury or damage to the works is

	excluded from any insurance required pursuant to clause 21.2: insurance – liability, etc. of employer.
22A.1, 22B.1, 22C.1 & 22C.2	The joint names insurance policies required by the applicable clause(s) has to be maintained by either the contractor or the employer, as the case may be, up to and including the date of issue of the certificate of practical completion (or the date of determina tion of the employment of the contractor if that is earlier).
22D.1	Where the appendix states that this clause applies the contractor has to maintain insurance for employer's loss of liquidated damages until the date of practical completion.
23.3.1	For the purposes of the works insurances the contractor retains possession of the site and the works up to and including the date of issue of the certificate of practical completion and the employer is not be entitled to take possession of the works until that date except as provided for in clause 18 (see later in this chapter).
23.3.2	Provided that the contractor gives his consent in writing and subject to the insurers confirming that insurance under clause 22A or 22B or 22C is not invalidated, the employer may use or occupy the site or the works for the storage of his goods or otherwise before the date of issue of the certificate of practical completion.
24.2.1	The employer is entitled to recover liquidated and ascertained damages from the contractor in respect of the period between the completion date and the date of practical completion.
25.3.3	Not later than the expiry of 12 weeks after the date of practical completion the architect has to fix a completion date later or earlier than previously fixed or confirm the completion date previously fixed.
27.2.1	If the contractor makes a specified default prior to the date of practical completion then, subject to giving the appropriate notices, the employer is entitled to determine the employment of the contractor.
28.2.2 to 28.2.5	If the carrying out of the works is suspended for a continuous period of the length stated in the appendix, prior to the date of practical completion, for one or more of the specified reasons then, subject to giving the appropriate notices, the contractor is entitled to determine the employment of the contractor.
28A.1	If the carrying out of the works is suspended for a continuous period of the length stated in the appendix, prior to the date of practical completion, for one or more of the specified reasons then, subject to giving the appropriate notices, either the employer or the contractor is entitled to determine the employment of the contractor.
30.1.3	Interim certificates have to be issued on the dates provided in the appendix up to the date of practical completion or to within one

month thereafter. Then interim certificates have to be issued as and when further amounts are ascertained as payable to the contractor.

30.4.1.2 & 30.4.1.3 In interim certificates the retention percentage may be deducted from the value of work which has not reached practical completion, but only half the retention percentage may be deducted from the value of work which has reached practical completion (until such time as the certificate of making good defects is issued).

30.6.1.1 Not later than 6 months after practical completion of the works the contractor has to provide, to the architect or to the quantity surveyor, all documents necessary for the purposes of the adjustment of the contract sum.

35.16 The architect must issue a certificate of practical completion for the works executed by each nominated sub-contractor.

35.19.1 Until the date of practical completion of the works (or the date when the employer takes possession of the works, if earlier) the contractor is responsible for loss or damage to any nominated sub-contract works for which a certificate of practical completion has been issued.

Sectional completion

Sectional completion arises when the employer requires the works to be completed in phased sections. In order to facilitate this procedure JCT98 is modified by the incorporation of the Sectional Completion Supplement, which includes a guidance note on its use (see also Chapter 10). In addition to setting out the necessary amendments to JCT98 the supplement contains a replacement appendix that differs from the JCT98 appendix as follows:

Clause	Difference
1.3	The date for completion of each section has to be stated, instead of a single date.
17.3	The defects liability period for each section has to be stated, instead of a single period.
18.1.5	An additional entry requiring the value of each section to be stated.
22D.2	The period of time for the insurance of employer's loss of liquidated damages for each section has to be stated, instead of a single period.
23.1.1	The date of possession for each section has to be stated, instead of a single date.
24.2.1	The rate of liquidated and ascertained damages for each section has to be stated, instead of a single rate.

The architect is required to issue a certificate of practical completion on the completion of each section. He is also required to issue a certificate of practical completion of the works as soon as practical completion of all the sections has been achieved.

Upon the issue of a certificate of practical completion of a section:

- the defects liability period comes into operation for the section
- defects, shrinkages or other faults caused by frost occurring before practical completion of the section, but not thereafter, have to be made good by the contractor at no cost to the employer
- if clause 19.1.2 is applicable, the employer's assignment rights become operable in respect of the section
- the section becomes 'property real or personal' to which the provisions of clause 20.2 apply
- the contractor or the employer is relieved of his duty to insure the section under clause 22A, 22B or 22C, whichever is applicable
- the contractor is relieved of his duty, if applicable, to maintain insurance for the employer's loss of liquidated damages in respect of the section
- the employer is entitled to recover liquidated and ascertained damages from the contractor in respect of the period between the completion date and the date of practical completion of the section
- the 12-week period commences during which the architect has to fix a completion date for that section later or earlier than previously fixed, or confirm the completion date for that section previously fixed.

If the project is large, it is quite feasible to have the situation where a certificate of practical completion is issued in respect of one section before work commences on other sections. Similarly the certificate of making good defects can be issued in respect of one section whilst the work remains incomplete on other sections.

Partial possession

As noted above, sectional completion arises when the employer requires the works to be completed in phased sections and the sectional completion supplement is incorporated in JCT98. Partial possession, however, arises out of a post-contract agreement between the employer and the contractor as provided for in JCT98 clause 18. If, before practical completion, the employer, with the agreement of the contractor, takes possession of a part of the works, the provisions of JCT98 clause 18 apply with the following effects:

- the architect has to immediately issue, to the contractor, a written statement identifying the part of the works taken into possession (the relevant part) and the date on which that possession occurred (the relevant date)
- in respect of the relevant part, practical completion is deemed to have occurred and the defects liability period is deemed to have commenced on the relevant date
- the architect has to issue a certificate of making good defects in respect of the relevant part when any defects, shrinkages or other faults in the relevant part, that he required to be made good, have been made good
- the contractor or the employer is relieved of his duty to insure the relevant part under clause 22A, 22B or 22C, whichever is applicable, but when clause 22C is applicable the employer will be obliged to insure the relevant part under clause 22C.1 from the relevant date
- the rate of liquidated and ascertained damages (LAD) stated in the appendix has to be revised as follows:

Revised LAD =

$$\frac{\text{Original LAD x (Contract sum − Value of Relevant Part in Contract sum)}}{\text{Contract sum}}$$

Possession of the building

A short time before practical completion the architect should ensure that the contractor has removed all of his plant, surplus materials, rubbish and temporary works from the site. When practical completion has been achieved and the architect has issued a certificate to that effect, the contractor ceases to be responsible for the works and the site and relinquishes possession of them to the employer. At this time the employer becomes responsible for all insurance matters relating to the site, the building(s) and contents.

At this stage it is advisable for the design team and the contractor to have a 'hand-over' meeting with the employer. During this meeting:

- the contractor can formally hand over to the employer all keys, properly labelled
- the contractor can hand over to the employer the building log-book, referred to in the Building Regulations 2000 Part L2 Section 3, giving details of the installed building services plant and controls, the method of operation and maintenance, and other details that collectively enable energy consumption to be monitored and controlled
- the employer and/or his staff can be fully briefed on the operation of the building and its services – when the building and/or its services are com-

plex this exercise may involve special training sessions for those who are to operate and maintain the facilities

- the contractor can hand over to the employer all relevant operating and maintenance manuals, guarantees and, where appropriate, a supply of spare parts for maintenance of critical equipment, etc.
- the design team and/or the contractor can hand over to the employer a full set of 'as built' drawings
- the planning supervisor can hand over to the employer the completed health and safety file.

Defects and making good

JCT98 clause 17.2 requires that not later than 14 days after the expiry of the defects liability period the architect has to prepare and deliver to the contractor as an instruction a schedule of all defects, shrinkages or other faults which are due to materials or workmanship not in accordance with the contract or to frost occurring before practical completion that have appeared during the defects liability period.

In practice it is usual for the employer to keep a record of such defects as they occur or are noticed and for this record to form the basis of the architect's schedule, which would normally be compiled during a thorough inspection of the works, by the design team, at the end of the defects liability period. Under normal circumstances the contractor has to make good all items on the schedule of defects at no cost to the employer and within a reasonable time of having received the instruction. However, JCT98 clause 17.2 does allow the architect, provided that he has the consent of the employer, to instruct the contractor not to make good some or all of the scheduled items. When the architect does so instruct, an appropriate deduction is made from the contract sum in respect of those items not to be made good.

If the architect requires any defect, shrinkage or fault to be made good during the defects liability period, JCT98 clause 17.3 permits him to so instruct the contractor, who has to comply with the instruction within a reasonable time. If the contractor fails to make good any item within a reasonable period the architect may, after having given the contractor the notice required by JCT98 clause 4, employ and pay others to execute the work and deduct all costs thereby incurred from monies due to the contractor. If there is insufficient money due to the contractor to fully defray the costs incurred, the shortfall is recoverable by the employer from the contractor as a debt.

When the architect is satisfied that all defects, shrinkages and other faults that he required to be made good have been made good, he has to issue a certificate to that effect. RIBA Publications has produced a pro forma certificate of completion of making good defects – a completed copy is given as Example 15/2 at the end of this chapter.

Final account

The responsibilities of the contractor, the architect and the quantity surveyor in connection with the final account are set out at the beginning of JCT98 clause 30.6 and are:

- not later than six months after practical completion of the works the contractor has to send to the architect or the quantity surveyor, if so instructed by the architect, all documents necessary for the adjustment of the contract sum
- within three months of the contractor sending all necessary documents, the architect or the quantity surveyor, if so instructed by the architect, has to ascertain any outstanding loss and/or expense and the quantity surveyor has to prepare a statement of all other adjustments to be made to the contract sum – the ascertainments of loss and/or expense and the statement of adjustments are, together, commonly referred to as the final account
- when complete, the architect has to send a copy of the final account to the contractor and copies of relevant extracts from it to each nominated sub-contractor.

Final account

It is worth noting that, under JCT98, the architect and the quantity surveyor have unilateral responsibility for the preparation of the final account – there is no requirement for them to obtain the contractor's agreement to their ascertainments and adjustments. However, in practice, it is normal for the consultants and the contractor to co-operate in reaching an agreed final account. When such agreement cannot be reached the contractor's only recourse is to invoke the contract dispute resolution procedures.

JCT98 clause 30.6 also sets out all of the matters that must be dealt with in the final account in order to properly adjust the contract sum in accordance with the conditions of contract. These are summarised below.

To be deducted or added as the case may be:
- the amount of any valuation agreed by the employer and the contractor for a variation
- the value of any accepted 13A quotation for a variation
- the value of any accepted price statement for a variation.

To be deducted:
- prime cost sums and amounts in respect of named nominated sub-contractors, together with any contractor's profit thereon, included in the contract bills
- any amounts included in interim certificates and paid by the employer in respect of any defective work by a nominated sub-contractor whose employment has been subsequently determined, together with any contractor's profit thereon
- all provisional sums and the value of all work included by way of approximate quantities in the contract bills
- the amount of the valuation of variations which are omissions
- the amount included in the contract bills for work which, due to a variation, has to be executed under substantially changed conditions
- any amount in respect of errors in setting out which the architect instructs are not to be amended
- any amount in respect of work not in accordance with the contract which the architect allows to remain
- any amount in respect of any defects, shrinkages or other faults which the architect instructs are not to be made good
- any amount allowable to the employer for fluctuations
- any other amount which is required, by the contract, to be deducted from the contract sum.

To be added:
- the amounts of the nominated sub-contract sums as finally adjusted or ascertained under the relevant provisions of conditions NSC/C

- the amount of the tender, adjusted in accordance with the terms of that tender, where the contractor has successfully tendered for work that was to have been carried out by a nominated sub-contractor
- the amounts properly payable by the employer for materials or goods obtained from nominated suppliers, including the discount for cash of 5% but excluding VAT
- the profit of the contractor on each of the three preceding items
- any amount payable by the employer in respect of statutory fees and charges, opening up work for inspection and testing, royalties arising out of compliance with an architect's instruction and insurance under clauses 21.2.3
- the amount of the valuation of variations, other than those which are omissions
- the amount of the valuation of any work which, due to a variation, has to be executed under substantially changed conditions
- the amount of the valuation of work executed as the result of instructions relating to the expenditure of provisional sums and all work included by way of approximate quantities in the contract bills
- any amount ascertained in respect of loss and/or expense arising from matters materially affecting the regular progress of the works or from the discovery of antiquities
- the amount of any costs incurred by the contractor if the employer defaults under clauses 22B or 22C
- any amount payable to the contractor for fluctuations
- any other amount which is required, by the contract, to be added to the contract sum.

It is good practice for each of the foregoing items to be identified separately in the final account and for separate amounts to be given for each variation, each nominated sub-contractor and each nominated supplier.

Whenever measurements have to be taken for the purpose of valuation the contractor must be given the opportunity of being present and of taking whatever notes and measurements he may require. It is to be noted that this facility is to be extended to the contractor both when measurements are to be taken on site, directly from the work, and when they are to be taken from drawings.

It is a worthwhile exercise to draft the final account as construction proceeds as it can be of considerable use to the building team for the purposes of:

- post-contract cost control and the preparation of regular financial reviews
- valuations for interim certificates
- speeding up the production of the final account once practical completion of the works has been achieved.

Whether or not this is done the design team must bear in mind the JCT98 requirement to complete the final account within three months of receipt from the contractor of all the documents necessary for its production. Any delays in doing so are likely to cause either the contractor or the employer to incur additional financing costs that are not recoverable under the contract.

Final certificate

JCT98 clause 30.8 requires the architect to issue the final certificate within two months of whichever of the following events occurs last:

- the end of the defects liability period, or
- the date of issue of the certificate of making good defects, or
- the date on which the architect sends a copy of the final account to the contractor.

The amount of the balance due either from the employer to the contractor or vice versa expressed in the final certificate is the difference between the total of the final account and the amounts previously stated as due in interim certificates plus the amount of any advance payments paid. The same clause also preserves the contractor's rights to any amounts previously included in interim certificates that have not been paid by the employer. RIBA Publications publishes a pro forma final certificate – a completed copy is given as Example 15/3 at the end of this chapter.

The JCT98 provision referred to in Chapter 14, which requires the architect to issue an interim certificate including the final amounts due to all nominated sub-contractors, not less than 28 days before the issue of the final certificate, means that the final certificate will not include any further amounts due to nominated sub-contractors. Nonetheless, the architect must notify each nominated sub-contractor of the date of issue of the final certificate.

JCT98 clause 30.9 prescribes the intended effect of the final certificate, which can be summarised as:

- conclusive evidence that where the particular quality of any materials etc. or standard of any workmanship is to be for the approval of the architect, such items are to the reasonable satisfaction of the architect
- conclusive evidence, save for accidental errors, that the contract sum has been properly adjusted in accordance with all the terms of the contract
- conclusive evidence that only those extensions of time that are due under the contract have been given
- conclusive evidence that the reimbursement of loss and/or expense is in full and final settlement of all and any claims that the contractor may have in respect of clause 26 matters.

Issued by: Ivor Barch Associates
address: Prospects Drive, Fairbridge

Certificate of
Practical Completion

Employer: Cosmeston Preparatory School
address: Fairbridge

JCT 98 / IFC 98

Job reference: IBA/94/20

Contractor: L&M Construction Ltd
address: Ferry Road, Fairbridge

Certificate no: 1

Issue date: 28 January 2000

Works: New School Library
situated at: Park Street, Fairbridge

Contract dated: 14 December 1998

Under the terms of the above-mentioned Contract,

I/we hereby certify that in my/our opinion

Practical Completion of

* the Works

*Delete as appropriate

* Section no 2 of the Works

has been achieved

SPECIMEN

* and the Contractor has complied with the contractual requirements in respect of information for the health and safety file

This item applies to JCT 98 only.

* and the Contractor has supplied the specified drawings and information relating to Performance Specified Work

on 28 January   20 00

To be signed by or for the issuer named above

Signed *Ivor Barch*

Distribution	Employer	☐	Structural Engineer	☐	Planning Supervisor	☐	☐	
	Contractor	☐	M&E Consultant	☐		☐	☐	
	Quantity Surveyor	☐	Clerk of Works	☐		☐	File	☐

F853A/B for JCT 98 / IFC 98

© RIBA Publications 1999

Example 15/1 Certificate of practical completion.

Issued by: Ivor Barch Associates
address: Prospects Drive, Fairbridge

Certificate of
Completion of

**Making Good
Defects**

Employer: Cosmeston Preparatory School
address: Fairbridge

JCT 98

Job reference: IBA/94/20

Contractor: L&M Construction Ltd
address: Ferry Road, Fairbridge

Certificate no: 1

Issue date: 8 September 2000

Works: New School Library
situated at: Park Street, Fairbridge

Contract dated: 14 December 1998

Under the terms of the above-mentioned Contract,

I/we hereby certify that the making good of

any defects, shrinkages or other faults specified in a schedule of defects
delivered to the Contractor as an instruction

and any other defect, shrinkage or fault which has appeared within the
Defects Liability Period and has been required by an instruction to be made
good

and relating to

*Delete as
appropriate

* the Works referred to in the Certificate of Practical Completion
no.   1   dated   28 January 2000

* Section no. _____ of the Works referred to in the
Certificate of Practical Completion
no. _____ dated _____

* the part of the Works identified in the
Statement of Partial Possession by the Employer

no.   1   dated   26 November 1999

was in my/our opinion completed on

6 September   20 00

To be signed by or for
the issuer named
above

Signed   *Ivor Barch*

Distribution

☐ Employer	☐ Structural Engineer	☐ Planning Supervisor	☐
☐ Contractor	☐ M&E Consultant	☐	☐
☐ Quantity Surveyor	☐ Clerk of Works	☐	☐ File

F807A for JCT 98   © RIBA Publications 1999

Example 15/2 Certificate of making good defects.

Issued by:	Ivor Barch Associates	**Final Certificate**
address:	Prospects Drive, Fairbridge	**JCT 98**
Employer:	Cosmeston Preparatory School	Serial no: **K**
address:	Fairbridge	Job reference: IBA/94/20
Contractor:	L&M Construction Ltd	Date of issue: 5 October 2000
address:	Ferry Road, Fairbridge	Final date for payment: 2 November 2000
Works: situated at:	New School Library Park Street, Fairbridge	

Contract dated: 14 December 1998

Original to Employer

This Final Certificate is issued under the terms of the above-mentioned Contract.

Contract Sum adjusted as necessary . . . . £   511,187.00

Total amount previously certified for payment to the Contractor plus amount of any advance payment . . . . £   502,064.00

Difference between the above stated amounts . . . . £   9,123.00

All amounts are exclusive of VAT.

I/We hereby certify the sum of (in words)

Nine Thousand One Hundred and Twenty Three Pounds only

as a **balance due**:

* *Delete as appropriate*

* to the Contractor from the Employer.

~~* to the Employer from the Contractor.~~

To be signed by or for the issuer named above

Signed *Ivor Barch*

[1] *Relevant only if clause 1A of the VAT Agreement applies. Delete if not applicable.*

[1] The Contractor has given notice that the rate of VAT chargeable on the supply of goods and services to which the Contract relates is   17.5   %

[1]   17.5   % of the amount certified above . . . . £   1,596.53

[1] Total of balance due and VAT amount (for information) . . . . £   10,719.53

This is not a Tax Invoice.

F852A for JCT 98   © RIBA Publications 1999

Example 15/3   Final certificate

Chapter 16
Delays and Disputes

Introduction

Delays to the progress of the works and the loss and/or expense associated with such delays are perhaps the most common causes of dispute encountered in the administration of building contracts. It is not surprising, therefore, that the draughtsmen of JCT98 paid particular attention to the procedures to be followed when delays occur, or are foreseen, as a result of which the contract period might become extended and/or the contractor may become entitled to the reimbursement of loss and/or expense.

If the employer, through default, prevents the contractor from completing the works by the contracted completion date and there are no express contractual provisions enabling the employer to set a new date for completion, the contractor's obligation is transformed from one of having to complete by the contracted date to one of completing in a reasonable time. In other words, time is 'at large' and the employer loses any right that he may otherwise have to levy liquidated and ascertained damages. It may be seen, therefore, that the existence and proper execution of an extension of time clause within a contract is of the utmost importance to the employer in that it preserves his other contractual rights should he or his representatives cause the works to be delayed by default.

Two of the definitions in JCT98 clause 1.3 that have particular significance in relation to the works being delayed are:

- *'Completion Date: the Date for Completion as fixed and stated in the Appendix or any date fixed either under clause 25 or in a confirmed acceptance of a 13A Quotation'*
- *'Relevant Event: any one of the events set out in clause 25.4'* (the events causing delay which may entitle the contractor to an extension of time).

All delays will fall into one of the following three categories:

- delays caused by the contractor, or
- delays caused by the employer or his representatives, or

- delays caused by events outside the control of both the contractor and the employer (often referred to in the industry as 'shared risk events' or 'neutral events').

Although only delays caused by the employer or his representatives or by neutral events are recognised by JCT98 clause 25.4 as relevant events which may give rise to an extension of time, clause 25.2.1.1 requires the contractor to give written notice to the architect whenever it becomes apparent that the progress of the works is being or is likely to be delayed, irrespective of cause.

Delays caused by the contractor

Within this category are all those delays that can be avoided if the contractor proceeds regularly and diligently with the works and uses his best endeavours at all times to prevent delays. JCT98 effectively requires the contractor to take all measures within his control to ensure:

- that there is an adequate labour force on the job
- that the necessary goods and materials are on site whenever they are needed
- that the works are not delayed by domestic sub-contractors.

If the contractor does not take all such measures and delays occur, there is no provision in JCT98 to enable the architect to grant an extension of time in respect of them. The responsibility for any such delays lies with the contractor and under JCT98 clause 24 he may well become liable to the employer for liquidated and ascertained damages in respect of the delay period.

If the contractor fails to proceed regularly and diligently with the works, the employer's ultimate sanction, subject to the giving of the required notices, is to determine the employment of the contractor under JCT98 clause 27.

Delays caused by the employer or his representatives

Certain of the relevant events recognised by JCT98 clause 25.4 result from the actions or inaction of the employer or his representatives. These are:

- * when an information release schedule has been provided by the employer, failure of the architect to release the information referred to in that schedule at the time stated in the schedule
- * failure of the architect to provide the contractor with the drawings, details and instructions to enable the contractor to carry out and complete the works in accordance with the conditions

- compliance with the architect's instructions to open up work for inspection or testing when the inspection or test shows that the work is in accordance with the contract
- *compliance with the architect's instructions in regard to any discrepancies within the contract documents
- * the execution of, or the failure to execute, work not forming part of the contract by the employer
- the supply of, or the failure to supply, materials or goods by the employer that the employer has agreed to provide
- * compliance with the architect's instructions in regard to the postponement of any work
- * failure of the employer to give ingress to or egress from the site of the works in due time
- compliance with the architect's instructions requiring a variation, except those for which a 13A quotation has been accepted
- compliance with the architect's instructions in regard to the expenditure of provisional sums, except those for defined work or for performance-specified work
- the execution of work for which an approximate quantity is included in the contract bills which is not a reasonably accurate forecast of the quantity of work required
- compliance or non-compliance by the employer with the requirements that both the planning supervisor and the principal contractor perform all of the duties required of them by CDM94
- suspension by the contractor of the performance of his obligations under the contract as a result of the employer's failure to pay a due amount in full
- any impediment, prevention or default by the employer or any person for whom the employer is responsible except to the extent that it was caused or contributed to by any default of the contractor or his servants, agents or sub-contractors.

Not only do all delays within this category entitle the contractor to an extension of time if they are likely to delay the completion of the works beyond the completion date; they may also cause the contractor to suffer loss and/or expense which, if incurred, and subject to the issue of the relevant notices, is reimbursable under the provisions of JCT98 clause 26.

If the employer fails to comply with CDM94 and/or if the carrying out of the works is suspended for a continuous period of the length stated in the appendix to JCT98 by reason of one or more of the above relevant events marked * or by reason of compliance with an architect's instruction requiring a variation, the contractor's ultimate sanction, subject to giving the required notices, is to determine the employment of the contractor under JCT98 clause 28.

Delays caused by events outside the control of the employer and the contractor

There are occasions when delays arise due to circumstances over which neither of the parties to the contract has any control. These neutral events, which are also recognised by JCT98 clause 25.4, are:

- force majeure
- exceptionally adverse weather conditions
- loss or damage caused by fire, lightning, explosion, storm, tempest, flood, bursting or overflowing of water tanks, apparatus or pipes, earthquake, aircraft and other aerial devices or articles dropped therefrom, riot or civil commotion
- local combination of workmen, strike or lockout affecting any of the trades employed on the works or any of the trades engaged in the preparation, manufacture or transportation of any of the goods or materials required for the works
- compliance with the architect's instructions in regard to any discrepancy or divergence between the contractor's statement in respect of performance-specified work and any instruction of the architect issued after receipt by the architect of the contractor's statement
- compliance with the architect's instructions in regard to what is to be done with any fossils, antiquities and other objects of interest or value that may be found on the site
- compliance with the architect's instructions in regard to nominated sub-contractors or nominated suppliers
- delay on the part of nominated sub-contractors or nominated suppliers which the contractor has taken all practicable steps to avoid or reduce
- exercise by the government of any statutory power which directly affects the execution of the works by restricting the availability or use of labour or by preventing or delaying the securing of goods, materials, fuel or energy
- the contractor's inability for reasons beyond his control and which could not reasonably have been foreseen at the contract base date to secure such labour, goods or materials as are essential to the proper execution of the works
- a local authority or statutory undertaker executing, or failing to execute, work in pursuance of its statutory obligations in relation to the works
- the deferment by the employer of giving possession of the site when clause 23.1.2 applies
- delay that the contractor has taken all reasonable steps to avoid or reduce, consequent upon a change in the statutory requirements after the base date, that necessitates some alteration or modification to any performance specified work

- the use or threat of terrorism and/or the activity of the relevant authority in dealing with such use or threat.

All delays within this category entitle the contractor to an extension of time if they are likely to delay the completion of the works beyond the completion date, but with the exception of those caused by compliance with the architect's instructions in regard to what is to be done with any fossils, antiquities and other objects of interest or value that may be found on the site, they do not entitle the contractor to reimbursement of any loss and/or expense incurred. In this way the parties share the burden of any delays caused by these neutral events. Reimbursement of any direct loss and/or expense arising from architect's instructions in regard to what is done with any fossils, antiquities etc. is covered by JCT98 clause 34 and not clause 26.

It is to be noted that in the case of delays on the part of nominated sub-contractors and delays consequent upon a change in the statutory requirements after the contract base date that necessitates some alteration or modification to any performance-specified work, JCT98 imposes an obligation on the contractor to take all practicable steps to avoid or reduce the delay. Should such delays occur, it is incumbent on the architect to satisfy himself that the contractor has taken such steps before he considers granting an extension of time.

Differences in the interpretation of certain of these neutral relevant events may give rise to disputes. It is therefore worth considering them in more detail.

Force majeure

This phrase comes from French law (Code Napoléon) and does not have a precise definition in English. Its literal translation is 'superior force' and in its broadest usage examples would include acts of God, war, strikes, weather conditions, epidemics, direct legislative or administrative interference, etc. However, several of these matters are provided for elsewhere in JCT98, hence force majeure is included as a general catch-all term to provide for any event or effect that cannot reasonably be anticipated or controlled and which is not otherwise expressly included in the contract.

Exceptionally adverse weather conditions

In earlier editions of the JCT standard forms of building contract the term 'exceptionally inclement weather' was used. The present term was introduced in the 1980 editions of the JCT contracts following the unusually long, hot summer of 1976, when many building projects were delayed by exceptional but not inclement weather.

'Exceptionally' is clearly the important word in this phrase and must be interpreted according to the time of year and the timing of the project anticipated by the contract documents. Thus if it were known at the time the contract were let that the execution of weather-sensitive work would span the winter period, then if the works were delayed by a two-week period of snow and frost during January such weather would most probably not be regarded as exceptionally adverse and would therefore not be a relevant event. However, should the snow and frost persist for a continuous period of eight weeks, this may well be regarded as being exceptionally adverse and therefore constitute a relevant event giving cause for an extension of time.

It is also important when interpreting the word 'exceptionally' to have regard to the location of the works, for what may be exceptional in the home counties may not be so in the north of England.

Delay on the part of nominated sub-contractors or nominated suppliers

Nominated sub-contractors and nominated suppliers are under the general control of the contractor but, because they are, in effect, imposed upon him by way of the employer's nomination, any delays which they may cause fall into this category of neutral relevant events so that the effects of the nominees' defaults are shared by both parties.

Delays on the part of nominated sub-contractors are often a cause of dispute, partly because of the difficulty in establishing precisely who is responsible for causing the delay and partly because the contractor is entitled, under

... delay on the part of nominated sub-contractors ...

JCT98, to an extension of time in respect of them provided that he is able to demonstrate that he has taken all practicable steps to avoid or reduce the delay. When such delays occur and the contractor is granted an extension of time, the employer is deprived of his means of redress for delay by way of liquidated and ascertained damages.

In the light of the above, the documentation used for the nomination of sub-contractors under JCT98 clause 35, that is Agreement NSC/W, establishes a direct contractual link between the employer and the nominated sub-contractor. Amongst other things this agreement provides that '*The sub-contractor shall so perform the Sub-Contract that the Main Contractor will not become entitled to an extension of time for completion of the Main Contract Works by reason of the Relevant Event in clause 25.4.7 of the Main Contract Conditions'.* Thus, if the nominated sub-contractor causes a delay to the main contract the employer will have an enforceable remedy against him without affecting the main contractor's entitlement to an extension of time.

JCT98 procedure in the event of delay

Clearly, it is in the interests of both the employer and the contractor to take all possible steps to avoid delays occurring, but when they do the architect must bear in mind that he may give extensions of time only in respect of delays caused by the relevant events listed in clause 25 and he may certify payment of loss and/or expense arising only from the matters listed in clauses 26 and 34. Any other delays or loss and/or expense that may be caused by the employer cannot be dealt with by the architect.

JCT98 clause 25.3.4 requires the contractor to '*use constantly his best endeavours to prevent delay in the progress of the Works and to prevent the completion of the Works being delayed or further delayed beyond the Completion Date'* as a condition precedent to the architect granting an extension of time. This is a demanding proviso and one which would seem to impose on the contractor an obligation to do everything possible to achieve the earlier completion date. It has been suggested that this obligation does not contemplate the expenditure of substantial sums of money. However, in using his best endeavours the contractor is expected to employ the same resources as would a prudent person to achieve his own ends and in doing so he will incur some cost. It is suggested that if the cause of delay is a JCT98 clause 26 matter, the reasonable cost to the contractor of using best endeavours is reimbursable as part of his loss and/or expense; otherwise it is not a recoverable cost. When the cost is not recoverable the contractor will have to decide the extent of the measures he is prepared to take in order to reduce his exposure to liquidated and ascertained damages in respect of any delays he may have caused.

It is worth noting that any delays caused by default of the employer that cannot be dealt with under the contract, for example failure to give possession of the site within the period of deferment stated in the appendix, can result in time being at large and the employer losing any right that he may otherwise have to levy liquidated and ascertained damages.

Whenever it becomes apparent that the progress of the works is being or is likely to be delayed, clause 25.2 requires the contractor to give written notice to the architect of the pertinent conditions and the event(s) causing delay. It also requires him, either in that notice or in a further written notice issued as soon after the first as possible, to identify those events that in his opinion are relevant and to give particulars of the expected effects and an estimate of the expected delay in the completion of the works in respect of each such event. These particulars and estimates have to be updated whenever necessary or as required by the architect. Copies of all notices required by clause 25.2 that include reference to a nominated sub-contractor have to be sent to the nominated sub-contractor.

If the architect is of the opinion that any of the notified events causing a delay is a relevant event and that the completion of the works is likely to be delayed beyond the completion date by reason of that delay, he has to fix a later completion date and notify the contractor, in writing, accordingly. RIBA Publications has published a pro forma notification of a revised completion date – a completed copy is given as Example 16/1 at the end of this chapter.

The architect should aim to either fix a new completion date or advise the contractor that it is not fair and reasonable to do so within 12 weeks of receiving an adequately detailed notice from the contractor or by the completion date if that is sooner, provided, of course, that the architect has, before that date, received an adequately detailed notice. If the architect fixes a new completion date he must state which of the relevant events he has taken into account and the extent, if any, to which he has had regard to variations requiring an omission of work.

If the architect issues any instructions after the first occasion on which he fixes a new completion date which result in the omission of work by way of:

- a variation, and/or
- the omission of a provisional sum for defined work, and/or
- the omission of a provisional sum for performance specified work

then he may fix a new earlier completion date, provided that, having regard to the work omitted, it is fair and reasonable to do so. However, when fixing an earlier completion date the architect is not allowed to alter the length of any extension of time required by the contractor and for which a 13A Quotation or the adjustment of time element of a Contractor's Price Statement has been accepted, nor is he allowed to fix a new completion date that is earlier than the date for completion stated in the appendix to the contract.

Not later than 12 weeks after the date of practical completion, the architect is required to finalise the position regarding the completion date by taking one of the following three courses of action.

(1) *Fix a completion date later than that previously fixed*

If, upon reviewing all of his previous decisions regarding extensions of time and upon reconsideration of all relevant events, whether or not specifically notified to him by the contractor, the architect reaches the conclusion that it is fair and reasonable to fix a completion date later than that previously fixed, he may do so. It will thus be seen that notification by the contractor of a relevant event is not, in the end, a condition precedent to him being granted an extension of time, although it is a condition precedent to the reparation of the completion date before *practical* completion.

(2) *Fix a completion date earlier than that previously fixed*

If, after the last occasion on which he fixed a new completion date, the architect issues instructions which result in the omission of work by way of a variation and/or the omission of a provisional sum for defined work and/or the omission of a provisional sum for performance-specified work and as a result of such omissions he reaches the conclusion that it is fair and reasonable to fix a completion date earlier than that previously fixed, he may do so. It is to be noted that fixing an earlier completion date has to be related to the omission of work; it cannot result from the architect reducing previously granted extensions of time which, with the benefit of hindsight, are thought to be overly generous. As when fixing an earlier completion date prior to practical completion, the architect is not allowed to alter the length of any extension of time required by the contractor and for which a 13A Quotation or the adjustment of time element of a Contractor's Price Statement has been accepted, nor is he allowed to fix a new completion date that is earlier than the date for completion stated in the appendix to the contract.

(3) *Confirm the completion date previously fixed*

Whichever course of action the architect takes, he must commit his decision to writing and forward it to the contractor with a copy to each nominated subcontractor.

There is little doubt that these procedures impose upon the contractor and the architect not inconsiderable obligations whenever there is a likelihood of delay. When seeking an extension of time the contractor must be specific and the architect must deal with the matter as and when it arises. The intention of JCT98 is quite clear – decisions on extensions of time are to be made as quickly as possible after the occurrence of a relevant event that is likely to cause a delay.

There is no place under JCT98 for the practice of waiting to see what the delay actually is at the end of the contract and then arguing about whether or not an extension of time is justified. Such practices, if followed, may well cause the contractor to incur additional costs in accelerating the works to meet an unnecessarily onerous completion date.

Whenever the architect has to review the completion date he should not overlook the fact that the master programme, which the contractor is obliged to provide and update so long as clause 5.3.1.2 has not been deleted, may well contain much valuable information to assist him in his deliberations.

Perhaps the greatest difficulty that the architect is likely to encounter in relation to the procedures to be followed in the event of delay is in deciding whether or not the notice, particulars and estimates that the contractor is required to provide are sufficient to make it practicable for him to fix a new completion date. In making this decision the architect should, as always, act fairly and must avoid the use of contrived allegations of insufficiency of notice, particulars and estimate as an excuse for delaying the fixing of a new completion date.

Reimbursement of loss and/or expense under JCT98

Any delay caused by default of the employer may give the contractor a right to claim damages at common law. As clause 26 sets down the procedures to be followed in the event of a fairly comprehensive list of matters causing disruption arising from the employer's acts or defaults, it is suggested that this clause might have restricted the contractor's common law entitlement had it not been for the provisions of clause 26.6 which states: *'The provisions of clause 26 are without prejudice to any other rights or remedies which the contractor may possess'.*

If the regular progress of the works is materially affected due to deferment of giving possession of the site (where clause 23.1.2 is applicable) or by one or more of the matters listed in clause 26.2, then subject to the timely issue of the relevant notices, the contractor is entitled to be reimbursed any direct loss and/or expense he may thereby incur and for which he would not be reimbursed by a payment under any other provision in the contract. These clause 26 matters are in fact the same as the relevant events, recognised by clause 25.4, which result from the actions or inaction of the employer or his representatives and which are listed earlier in this chapter under the heading 'Delays caused by the employer or his representatives'.

If the contractor considers that he has incurred or is likely to incur direct loss and/or expense due to deferment or as the result of one of the clause 26 matters, he must make written application to the architect as soon as it becomes apparent to him that the regular progress of the works has been, or is likely to be, materially affected. In such circumstances the contractor should give the architect as much information as he can but, when required to do so, he must submit to the architect such information as should enable him to form an opinion as to whether or not direct loss and/or expense has been incurred as a result of the matters cited in the contractor's application. Also, upon request, the contractor has to submit to the architect or to the quantity surveyor such details as are reasonably necessary to enable them to ascertain the amount of loss and/or expense which has been or which is being incurred. Any amount so ascertained must be included in the next interim certificate and is not subject to a deduction for retention (see Chapter 14).

It must always be borne in mind that there can be an entitlement to an extension of time without an entitlement to reimbursement of loss and/or expense. Thus, although the contractor may be granted an extension of time as a result of a delay caused by one of the clause 25 relevant events, he is not entitled to reimbursement of direct loss and/or expense unless he can show that the cause of the delay was also a clause 26 matter and that it actually caused him to suffer direct loss and/or expense which he could not recover under any other provision in the contract, for example through the valuation of a variation. Conversely there can be an entitlement to reimbursement of loss and/or expense without an entitlement to an extension of time. This can occur when the delay caused by the relevant event is not likely to affect the completion date but the clause 26 matter does cause the contractor to suffer loss and/or expense, for example delay to a non-critical element of the works, which causes additional expense due to loss of output.

Any direct loss and/or expense which in the opinion of the architect is incurred by the contractor as a result of compliance with the contractual provisions relating to or with architect's instructions issued in regard to what is to be done concerning any fossils, antiquities and other objects of interest or value found on the site, is reimbursable pursuant to clause 34. Whilst the clause 34 provisions relating to the ascertainment and reimbursement of direct loss and/or expense are very similar to those in clause 26, they do not, unlike clause 26, make ascertainment and reimbursement conditional upon the contractor making written application to the architect.

Liquidated and ascertained damages

It is normal in building contracts to provide for the employer to recover damages from the contractor if he fails to complete the works by the completion date. To avoid such damages being set aside by the courts as a penalty they must be liquidated and ascertained, that is, they must be a genuine pre-estimate of the likely loss that will be suffered by the employer if there is a delay to the completion of the works.

JCT98 makes provision for the recovery by the employer of liquidated and ascertained damages in clause 24. It also requires the insertion of an amount and period for those damages in the appendix against clause 24.2. It is important that the entry in the appendix is carefully considered. The amount of liquidated and ascertained damages applies to the whole of the works and should not therefore be expressed in terms of £/dwelling or £/unit. The period should not be stated as 'per week or part thereof' as this may, in the event, include a number of days after the date of practical completion – this could be held to be punitive and thus render the liquidated and ascertained damages unenforceable. To avoid this potential problem it is recommended that the period be stated as being 'per calendar week or pro rata thereto'.

The operation of JCT98 clause 24 is summarised below:

- If the contractor fails to complete the works by the completion date the architect is required to issue a certificate to that effect.
- Provided that such a certificate has been issued, the employer may, if he so wishes, inform the contractor in writing that he may withhold, deduct or require payment of liquidated and ascertained damages. If the employer chooses to so inform the contractor, he must do so before the date of the final certificate.
- The employer may then advise the contractor in writing that he is to pay the employer liquidated and ascertained damages at the rate stated in the appendix, or at a lesser rate if the employer so chooses, for the period between the completion date and the date of practical completion. Alternatively the employer may advise the contractor in writing that such liquidated and ascertained damages will be deducted from monies due to him.
- If, after the payment or withholding of liquidated and ascertained damages, the architect fixes a later completion date, the employer must pay or repay to the contractor any amounts recovered, allowed or paid in respect of the period between the two completion dates.

Disputes and dispute resolution

As a consequence of building projects in general being complicated, unique ventures erected largely in the open, on ground the condition of which is never fully predictable and in weather conditions that are even less so, it is not uncommon for disputes to arise during the course of building operations. There are many reasons why disputes occur, but in the main they are caused by the failure of one or more members of the building team:

- to do their work correctly, efficiently and in a timely manner
- to express themselves clearly, or
- to understand the full implications of instructions given or received.

Most disputes are of a minor nature and are settled quickly, fairly and amicably by the building team. From time to time, however, more serious issues come into dispute. When this happens, the building team should make every effort to reach a fair settlement by negotiation. If this fails, it becomes necessary to use one or more of the dispute resolution mechanisms available – mediation, adjudication, arbitration and litigation. In building contracts arbitration was hitherto the one used most commonly, but since the Housing Grants Construction and Regeneration Act 1996 (Construction Act) came into force on 1 May 1998, it has been overtaken by adjudication.

Mediation

Mediation is a private, informal process in which one or more neutral parties assist the disputants in their efforts towards settlement. As with negotiation, the resolution lies ultimately with those in dispute; the mediator acts purely as a facilitator and cannot impose his decision on the parties.

Adjudication

Until 1 May 1998 the chief weakness of adjudication in the construction industry was that it was largely unenforceable when push came to shove. However, under the Construction Act, adjudication is now the single most common mechanism by which construction disputes are resolved. Most building contracts now contain adjudication provisions compliant with the Construction Act. Those that do not are caught instead by statutory adjudication that gives a party to a building contract the unilateral right to refer a dispute arising under the building contract to adjudication.

Adjudication is a binding process, available to the parties at any time, in which the party referring the dispute will receive a decision within 28 days of the referral irrespective of the complexity of the issues in dispute or the amount of money at stake. The 28-day period can be extended by 14 days if the adjudicator obtains the express approval of the referring party, but it can be further extended only if both parties so agree. The decision is binding until the dispute is finally determined by arbitration or legal proceedings or by agreement. The parties may, however, agree to accept the decision of the adjudicator as finally determining the dispute. Even if the adjudicator makes a mistake in his decision, perhaps as a result of the sheer speed of the process, the courts will readily uphold it, leaving the disaffected party to raise separate proceedings in arbitration or litigation.

JCT98 Article 5 provides for any dispute or difference arising under the contract to be referred to adjudication by either party and clause 41A sets out, in compliance with the Construction Act, the detailed provisions for adjudication. These can be summarised as follows:

- The adjudicator is an individual, who is prepared to execute the JCT Adjudication Agreement, and is either agreed by the parties or is named by the nominator given in the appendix to the contract.
- The parties must act with the objective of securing the appointment of an adjudicator and the referral of the dispute or difference to the adjudicator within seven days of either party having given notice of his intention to refer a dispute or difference to adjudication.
- The parties are required to execute, with the adjudicator, the JCT Adjudication Agreement. However, if either party fails to do so it does not invalidate the decision of the adjudicator.
- If the adjudicator dies or becomes ill or is otherwise unavailable to adjudicate on the matter referred to him, the parties may agree upon an individual to replace him or either party may apply to the nominator for the nomination of a replacement.
- If the adjudicator is agreed or appointed within seven days of the notice to refer, the party giving the notice must refer the dispute or difference to the adjudicator within seven days of the notice; otherwise the referral has to be made immediately upon the agreement or appointment of the adjudicator. Particulars of the dispute or difference together with a summary of contentions relied upon, a statement of the relief or remedy sought and any other material the referring party wishes the adjudicator to consider are to accompany the referral. The referral and accompanying documents are to be copied simultaneously to the other party.
- If the referral and accompanying documents are sent to the adjudicator and/or the other party by fax then, for record purposes, further copies must be sent immediately by first-class post or by actual delivery.
- The adjudicator is required to confirm receipt of the referral and accompanying documents.
- Within seven days of the date of the referral, the party not making the referral may send to the adjudicator, with a copy to the other party, a written statement of the contentions on which he relies, together with any other material that he wishes the adjudicator to consider.
- Within 28 days of the referral, or such longer period as may be agreed by the parties (see above), the adjudicator has to send his decision, in writing, to the parties. In reaching his decision the adjudicator must act as such and not as an expert or an arbitrator. He is not obliged to give reasons for his decision.
- The adjudicator is required to act impartially. He may set his own procedure and he has absolute discretion to take the initiative in ascertaining the facts and the law. He may also:
  - use his own knowledge and experience
  - open up, review and revise any certificate, opinion, decision, requirement or notice issued under the contract
  - require the parties to provide further information

- require the parties to carry out tests and to open up work
- visit the site or any workshop where work is or has been prepared for the contract
- obtain any information he considers necessary from any employee or representative of the parties provided he gives the party concerned prior notice
- obtain such technical or legal advice from others as he considers necessary, provided that he gives the parties prior notice and a statement or estimate of the cost involved
- decide the circumstances in which, or the period for which, a simple rate for interest has to be paid.
- The parties have to meet their own costs but the adjudicator can direct as to who is to pay the cost of any tests or opening up that he may have required.
- The parties are jointly and severally liable for the payment of the adjudicator's fee and reasonable expenses. The adjudicator may direct how his fee and expenses are to be apportioned between the parties. If he does not do so they are borne equally by the parties.
- The decision of the adjudicator is binding (see above) and the parties are required to comply with the decision. Should either party not comply, the other party is entitled to take legal proceedings to secure such compliance.
- The adjudicator is not liable for anything he does or omits to do in the discharge of his functions as adjudicator unless he acts in bad faith.

Arbitration

Arbitration is a process, subject to statutory controls, whereby a private tribunal of the parties' choosing determines a formal dispute.

Many people in the construction industry still take the view that it is essential to have their disputes heard by someone with a working knowledge of building contracts and the practices that are prevalent in the industry. They often lean towards arbitration as a means of achieving that end, but they frequently fail to take account of the expertise available within the Technology and Construction Court (TCC), which is almost invariably the court, within the High Court, that hears building contract matters.

Arbitration, therefore, essentially represents a process that is available as an alternative to litigation with the purported advantages of flexibility, economy, expedition, privacy and freedom of choice.

JCT98 Article 7A provides for either party to refer to arbitration any dispute or difference, arising under the contract either during the progress or after the completion or abandonment of the works or after the determination of the employment of the contractor, except:

- for those in connection with the enforcement of any decision of an adjudicator
- for those in connection with the Construction Industry Scheme where any Act of Parliament or statutory instrument provides for some other method of resolving the dispute or difference
- for those in connection with the employer's right to challenge VAT claimed by the contractor.

This provision is an alternative to Article 7B which provides for the settlement of disputes or differences by legal proceedings. The applicable alternative is set down in the appendix to the contract. When arbitration applies, the relevant provisions are to be found in clause 41B and can be summarised as follows:

- The provisions of the Arbitration Act 1996 apply to any arbitration under the contract and any such arbitration has to be conducted in accordance with the JCT 1998 edition of the Construction Industry Model Arbitration Rules (CIMAR) current at the contract base date.
- If either party requires a dispute or difference to be referred to arbitration then that party has to serve on the other party a written notice of arbitration identifying the dispute and requiring him to agree to the appointment of an arbitrator.
- If the parties fail to agree on the name of an arbitrator within 14 days of the service of the notice of arbitration then either party may apply to the appointor, named in the appendix to the contract, for the appointment of an arbitrator.
- If the previously appointed arbitrator ceases to hold office, for any reason, the parties may agree upon an individual to replace him or either party may apply to the appointor for the appointment of a replacement.
- Where two or more related arbitral proceedings fall under separate arbitration agreements, the persons who are to appoint the arbitrator may appoint the same arbitrator for all related proceedings.
- After an arbitrator has been appointed either party may give notice to the other and to the arbitrator referring any other dispute under the contract to be decided in the arbitral proceedings.
- The arbitrator has to determine all matters in dispute that are submitted to him and in doing so he has the power:
  - to rectify the contract so that it accurately reflects the true agreement made by the parties
  - to require such measurements and valuations as he considers necessary to determine the rights of the parties
  - to ascertain and award any sum that ought to have been included in any certificate
  - to open up, review and revise any certificate, opinion, decision, requirement or notice issued under the contract.

- Either party is at liberty to apply to the courts to determine a question of law arising in the course of a reference.
- Subject only to the right of either party to appeal to the courts, any question of law arising out of an award of the arbitrator is final and binding.

Litigation

The courts provide the setting for the traditional mode of dispute resolution – namely litigation. The number of disputes actually determined by the courts is negligible compared with the number of disputes settled by other means. Furthermore, very few of the proceedings that are begun actually result in a full trial and subsequent judgment. In fact, more than 90% of actions begun in the High Court are disposed of before reaching trial.

Most building disputes are dealt with by what were previously known as the official referees, now known as the technology and construction court judges who are, almost without exception, appointed from leading counsel and spend a great deal of their time hearing disputes about building contracts.

In his 1996 report 'Access to Justice', Lord Woolf expressed the opinion that the then current system of litigation was too expensive, too slow, too fragmented, too adversarial, too uncertain and incomprehensible to most litigants. As a result of his review, the Civil Procedure Rules were implemented on 26 April 1999 and apply to all actions begun from that date. It is worth just a mention here that a key facet of these rules, the Overriding Objective, was designed to help ensure that all cases are dealt with justly and in ways that are proportionate to the amount of money involved, the importance of the case, the complexity of the issue and the parties' financial position. There are now three tracks on which litigation runs:

- the small claims track for disputes of less than £5,000
- the fast track for disputes between £5,000 and £15,000
- the multi-track for disputes over £15,000.

JCT98 Article 7B provides for any dispute or difference, arising under the contract either during the progress or after the completion or abandonment of the works or after the determination of the employment of the contractor, to be determined by legal proceedings. This provision is an alternative to Article 7A which provides for the settlement of disputes or differences by arbitration. As noted above, the applicable alternative is set down in the appendix to the contract.

Notification of
Revision to

Completion Date

| Issued by: | Ivor Barch Associates |
| address: | Prospects Drive, Fairbridge |

JCT 98

| Employer: | Cosmeston Preparatory School |
| address: | Fairbridge |

Job reference: IBA/94/20

| Contractor: | L&M Construction Ltd |
| address: | Ferry Road, Fairbridge |

Notification no: 6

Issue date: 6 January 2000

| Works: | New School Library |
| situated at: | Park Street, Fairbridge |

Contract dated: 14 December 1998

Under the terms of the above-mentioned Contract,

I/we give notice that the Completion Date for

*Delete as appropriate

* the Works
~~* Section no.~~——————————~~of the Works~~

previously fixed as

17 December ~~20~~ 1999

* is hereby fixed later than that previously fixed,
~~* is hereby fixed earlier than that previously fixed,~~
~~* is hereby confirmed.~~

and is now

21 January 20 00

* This revision has taken into account the following Relevant Events:
  Exceptionally adverse weather conditions (cl. 25.4.2)
  Compliance with the Architect's instructions nos. 46
  and 47 (cl. 25.4.5.1)
~~* This revision has taken into account the omission of work required by the following Instructions:~~

~~* This revision is made by reason of my/our review.~~

To be signed by or for the issuer named above

Signed *Ivor Barch*

Distribution			
☐ Contractor	☐ Quantity Surveyor	☐ Clerk of Works	☐
☐ Employer	☐ Structural Engineer	☐ Planning Supervisor	☐
☐ Nominated Sub-Contractors	☐ M&E Consultant	☐	☐ File

F808 for JCT 98

© RIBA Publications 1999

Example 16/1   Revision to completion date.

Chapter 17
Capital Allowances

Introduction

Capital allowances are a form of tax relief on commercial property that are among the most significant and yet most frequently underutilised financial incentives in the UK. A series of tax breaks, they are intended to encourage businesses to invest in capital equipment and commercial property.

Almost all commercial property owners and users who incur capital expenditure on property are entitled to claim capital allowances. As much as 40% of construction costs and up to 75% of fitting-out costs can be allowable, releasing much-needed capital into the business itself. However, research indicates that about half of all businesses miss out on such opportunities. This is no doubt because the issues are complex and require a specialist who is expert in both property costs and capital allowances legislation in order to identify, value and negotiate capital allowances with the Inland Revenue.

The main types of capital allowances are:

- **plant and machinery allowances** – see the questions and answers at the end of this chapter for details of what may qualify for relief
- **industrial buildings allowances** – relief is given on the entire amount of capital expenditure providing the unit is used for a qualifying trade, as defined in the current Capital Allowances Act
- **hotel allowances** – relief is given on the entire capital expenditure providing the property is a qualifying hotel, as set out in the current Capital Allowances Act
- **research and development allowances** – qualifying trades (generally businesses involved in technological development) can claim for property-related research and development costs.

When capital allowances apply

The following property transactions or construction-related activities are usually triggers for capital allowances:

- purchase of an existing property
- refurbishment of existing premises by landlords, owner-occupiers or tenants
- new build, alterations and extensions by tenants/leaseholders or freeholders.

The worth of capital allowances

The actual cash benefit depends on the marginal tax rate of the business. Every £100,000 of plant and machinery claimed by a business will mean a tax saving of £30,000 over time, assuming a 30% corporation tax rate. Similarly, a higher rate income tax payer would save proportionately more. The total value of plant and machinery is not all claimed in the first year but at 25% per annum on a reducing balance basis. Since July 1998 the first year's allowance has been available at 40% for some small and medium-sized enterprises, but this reverts to 25% per annum in the second and subsequent years, again on a reducing balance basis. Industrial building allowances and hotel allowances are claimable at 4% per annum from the date of first use until they are exhausted after 25 years. Research and development allowances are all available in the year of expenditure.

Different types of properties qualify businesses for different levels of allowances. The following are typical percentages, reflecting the proportion of total construction cost accounted for by plant and machinery:

- Air conditioned offices – 25% to 40%.
- Heated offices – 17% to 25%.
- Hotels – 30% to 50% (plus hotel allowances).
- Industrial properties – 5% to12% (plus industrial building allowances).
- Fitting-out contracts – 40% to 75%.

Eligibility for capital allowances

The claimant must be liable to income tax or corporation tax on profits arising from their business activity. The statute does, however, require further criteria to be fulfilled. To qualify for plant and machinery allowances, a tax payer must incur capital expenditure upon the provision of plant and machinery that is owned by him and is used in the course of his trade.

Claimants tend to be either property investors or owner-occupiers who have incurred capital expenditure on the construction, refurbishment or fitting-out of their premises for the purpose of their business. The basis of any claim is therefore an analysis of actual capital expenditure incurred plus a proportion of associated professional fees.

When second-hand property is purchased the allowances usually attach to the title (freehold or leasehold) on which the original expenditure was incurred. The basis of any claim in these situations is different. The purchase price has to be split into its asset components:

- land cost
- building cost
- the cost of fixed plant and machinery within the building.

Valuations have to be obtained for each of the three components and apportioned using Inland Revenue guidelines to equal the price actually paid for the entire asset.

Tax planning to optimise capital allowances

Tax planning is a pre-requisite to maximising a tax payer's entitlement to capital allowances. Ultimately, the aim of the exercise is to compile documentary evidence and use precedents in case law to strengthen and support the claim. By following the design and construction processes and recording why particular design decisions are made the business use test can be overcome and the value of a claim can be increased by an average of 20%–25% on prestige and other complex new buildings, refurbishment and fitting-out contracts.

It is not always the nature of the item itself that determines eligibility for capital allowances. The way particular features or building elements function in the user's trade will govern whether an item passes or fails the business use test. The following list is typical of the items on which a tax planning exercise would focus:

It is not always the nature of the item itself that determines eligibility ...

- builders' work in connection with services
- powered doors or shutters
- built-in furniture
- cold water services installation
- above-ground drainage installation
- electrical installation
- raised demountable floors
- suspended ceilings, other than plenum ceilings
- fire officers requirements in existing buildings
- demountable partitions
- preliminaries, professional fees and other project 'on-costs'.

Historical expenditure

If historical expenditure has been overlooked, allowances are not lost; they can be deferred to a later period, provided that the asset is still owned. Under self-assessment it is possible to insert unclaimed allowances in an earlier period through an amended tax return. This is however subject to a time limit which, at the time of writing, was 12 months from the filing date for the period in which the allowances are being claimed.

Why capital allowances are not claimed

As noted above, only about 50% of owners or occupiers actually claim their proper entitlement to fixed plant and machinery allowances. The likely reasons why the rest do not are:

- poor cost information preventing a detailed analysis into 'qualifying' categories
- no appreciation of how allowances are valued on second-hand property
- mistakenly considering the exercise to be a tax issue when in fact it relates to issues of property cost and apportionment
- no access to proper specialist advice
- leaving the exercise too late (after practical completion on construction work and after the completion date on purchases).

How to be tax efficient

The employer should:

- always consider capital allowances early

- obtain a capital allowances forecast during feasibility stages and factor this into the financial appraisal
- obtain a forecast of the capital allowances before exchanging on purchased property to determine the increased effects on yields
- on larger schemes consider a tax-planning design audit
- never disregard historical expenditure.

Some questions and answers

Q. Can I get capital allowances for any of my fitting-out works?

A. Yes. There could be a significant level of capital allowances available as a result of incurring capital expenditure on fitting-out works. It is not uncommon to find that as much as 75% of such expenditure qualifies for plant and machinery allowances.

Q. What is plant and machinery?

A. The Capital Allowances Act precludes most parts of buildings and permanent structures from qualifying for capital allowances as plant and machinery. Items that could qualify include air conditioning, heating, lifts, electrical installations (either in part or in their entirety, depending on the circumstances), sanitary ware, hot water, carpets, blinds, demountable partitioning, signage, and certain types of access flooring systems and suspended ceilings. In retail stores, branch offices and similar premises, large sums of money are often spent on achieving the corporate image. In hotels, public houses and restaurants, money is spent on decorative assets to create the right atmosphere or ambience. Where it can be shown that these items pass what the courts have described as the 'business use' test and the 'premises' test, the expenditure on them should qualify for capital allowances. It is therefore important to bear in mind that it is not just the item but also the reason why it was installed that is important.

Q. I do not have a breakdown of the fitting-out costs; can I still claim?

A. Yes. Problems often occur with lump-sum contracts priced on the basis of a specification and drawings. If, however, the total cost is broken down using traditional surveying skills and then documented, a claim can be made on the identified qualifying expenditure.

Q. Does self-assessment affect capital allowances?

A. Yes. It affects the way claims are prepared. Previously a small minority often submitted percentage claims with little or no back-up in support of the figures. Self-assessment has introduced new obligations for both tax payers and advisers with penalties for non-compliance. This means being able to produce a complete audit trail whether or not the Inland Revenue actually calls for it.

Q. What about capital allowances on refurbishment schemes?

A. Capital expenditure on these types of projects will attract high levels of capital allowances. The key area to consider is works incidental to the installation of plant and machinery. The Capital Allowances Act allows expenditure on alterations to existing buildings which are incidental to the installation of plant and machinery to be treated as part of the qualifying expenditure. In addition, some of the expenditure may be reclassified as a revenue expense, the whole of which would, in consequence, be deductible in the year in which it was incurred.

Q. Does the building design affect the level of capital allowances?

A. Yes. The more complex a building design, the greater the scope for tax planning capital allowances. Specially designed, newly constructed, owner-occupied properties can qualify for capital allowances as high as 40% of total expenditure. Although it may not always make long-term economic sense to choose design and engineering solutions purely because they attract capital allowances, it is worth commissioning a design 'tax' audit. Then, when decisions are made, the tax consequences are known. On projects worth many millions of pounds, even small changes, initiated at an early enough stage, can have a major influence on the ultimate eligibility of items for capital allowances.

Chapter 18
Insolvency

Introduction

In simple terms, insolvency means being unable to pay debts as and when they become due. It is a liquidity or cash flow problem which can affect even wealthy companies in difficult trading conditions. This chapter deals with corporate insolvency under building contracts, rather than bankruptcy, which is individual insolvency applying only to sole traders and partnerships.

Unlike many other events which occur during the course of a contract and can be planned for, insolvency is one over which the building team has no control and thus the procedure consequent upon insolvency can be actioned

"You never told me you'd put the DEBTS in my name."

HM PRISON VISITORS THIS SIDE

only after the event. While insolvency can overtake either party to a contract, generally speaking it happens more often to the contractor than to the employer. This chapter therefore concentrates on contractors' insolvency, but we look briefly at insolvency of employers and of nominated sub-contractors at the end.

Before considering the procedure which is laid down within JCT98, it is important to have an understanding of insolvency generally, in order to put that procedure in its proper context. It is also important to realise that the contractual provisions must be interpreted within general insolvency law, which can be complicated, and the parties may therefore need to seek expert legal advice.

Liquidation

Liquidation or winding up of a company is a legal process that brings its existence to an end in order that its assets can be realised for the benefit of its creditors and members. There are two types of liquidation in respect of companies: voluntary and compulsory.

Voluntary liquidation arises in two situations. A members' voluntary liquidation occurs when the members or shareholders of the company at an extraordinary general meeting resolve to wind up the company, which is still warranted by the directors to be solvent. However, if the company is considered to be insolvent, a creditors' meeting will follow the shareholders' meeting so that a liquidator can be appointed and in this case it becomes a creditors' voluntary liquidation.

Compulsory liquidation arises when the company or one or more creditors petition the court for the company to be wound up or placed in liquidation. In this case, if the court considers that the company is unable to meet its debts, it will order the winding up. Then a provisional liquidator, usually the official receiver, is appointed by the court pending the appointment of an official liquidator by the creditors at a meeting. It is possible for the official receiver to be appointed to the post, but it is more common for an independent insolvency practitioner to be appointed, except in 'no asset' cases when the official receiver continues to act because there are no assets to pay a private practitioner.

Insolvency practitioners

Insolvency practitioners are licensed by professional bodies and, since the Insolvency Act 1986, any person acting as a liquidator, provisional liquidator, administrative receiver, administrator or nominee/supervisor of a voluntary arrangement must be an insolvency practitioner.

Receivership

Receivership, which is governed by the Insolvency Act 1986, can be ordinary or administrative and is a process whereby an insolvency practitioner is appointed under specific instructions, which need not necessarily end in liquidation. A receiver will usually achieve a better realisation of the assets than a liquidator, being able to trade the asset(s) and sell the goodwill, thus obtaining the best value possible, consistent with the time and cost involved in achieving the realisation.

In ordinary receivership, a receiver is appointed by a debenture holder, often a bank, under the express provisions of the debenture (see below). His job is to realise the asset(s) over which he has control to satisfy the debt due to the appointor, after which he withdraws and any surplus is handed back to the directors. This type of receivership may lead to liquidation if the residue of the company is incapable of trading.

In administrative receivership, the administrative receiver is appointed by the holder of a debenture secured by a floating charge (typically a bank or other lending institution). The appointment has immediate effect, there being no waiting period pending a court decision or meeting of creditors, and imposes an immediate stop on payments by the company to its creditors. The administrative receiver's powers are derived from the debenture and the Insolvency Act 1986 and involve taking control of the whole company (not just a particular asset). The principal duty of the administrative receiver is to realise the assets charged by the debenture for the benefit of the secured creditor, but only after the preferential creditors (those with a fixed charge and the statutory entitlements), who have legal priority, have been satisfied. Any surplus funds realised will be returned to the directors of the company or to the liquidator, if appointed, for distribution to the unsecured creditors and shareholders of the company.

Administration

Administration is a procedure instigated by the Insolvency Act 1986, whereby an administrator is appointed by the court following an application by the company or a creditor. There are two prerequisites for the granting of an administration order – first, that the company is, or is likely to become, insolvent, and second, that the creditors' interests will be served thereby. Although the agreement of any debenture holders is not required, in practice they have to be persuaded, otherwise they can individually enforce their own security and appoint an administrative receiver, forestalling the court appointment. The purpose of the court appointment is to protect the company from its creditors so that a rescue package can be put in place – a moratorium being placed on the payment of the creditors. The role of the administrator is to keep the company functioning as a going concern and, hopefully, to trade the company out of its financial difficulties within a given timescale. If unsuccessful, the company may eventually have to go into liquidation.

An administrator hopes to preserve the assets of the company while satisfying the creditors, but where realisation has to take place, he will usually achieve a better realisation than a liquidator or receiver, as he will often restructure the business prior to disposal, obtaining the best value possible, consistent with the time and cost involved in achieving the realisation.

Administration has not been as successful as the government originally hoped in setting up the scheme under the 1986 Act – it is often frustrated by debenture holders objecting to the appointment of an administrator, and is expensive to operate due to the high setting-up and running costs.

Voluntary arrangements and compositions

When a company runs into cash flow difficulties, there is usually a period before insolvency is reached when, by planning cash flow on the basis of extended credit, a company can work itself out of its difficulties and back into normal trading. The formalisation of these extended credit facilities is called an arrangement for a composition of debts or a scheme of arrangement under the Companies Act 1985 (amended by the Companies Act 1989) or the Insolvency Act 1986. These Acts give statutory backing to such rescue arrangements after a strict procedure has been followed.

The procedure is initiated by the company proposing the arrangement under the supervision of an insolvency practitioner, called a 'nominee'. The nominee must prepare a report for the court and will decide whether a meeting of the company and creditors should be called (and the arrangements therefore). In order to be implemented, the proposal must have the support of 75% in value of the creditors. Once approved the nominee becomes the 'supervisor' of the arrangement.

It should be noted that the Acts do not give an immediate moratorium on creditor demands and so the process can be undermined by any individual creditor pursuing its individual rights before the arrangement is agreed. Further, such arrangements are effective only against creditors who have received notice of the proposed arrangement, and a creditor having a floating charge debenture maintains the right to appoint an administrative receiver at any time, in any event.

Debenture

A debenture is a bond, evidencing a debt, which contains conditions as to interest payable and as to its settlement, discharge or satisfaction. A debenture is generally secured by a fixed or floating charge on a company's assets, registered at Companies House. A fixed charge identifies a particular asset which can be liquidated to pay the debt; a floating charge is a charge on the general assets of the company. When the asset subject to a fixed charge is a building, the debenture is commonly arranged as a mortgage.

Fixed and floating charges rank as secured, or preferential, creditors in a winding up. Fixed charges rank first, followed by statutory responsibilities, such as unpaid wages, PAYE and VAT, and then floating charges. The unsecured creditors are, of course, bottom of the list and often receive nothing at all.

The procedure upon the insolvency of the contractor

It is important to understand that the procedure is carefully codified within JCT98 and the detailed wording of clause 27 of the contract should be studied in addition to the general advice given in this chapter. The Joint Contracts Tribunal has also published Practice Note 24, giving a detailed commentary on the contract clauses and the procedure involved, which is extremely valuable.

Clause 27 draws a distinction between those insolvency procedures which are orientated towards business rescue and those which lead to the inevitable demise of the company. In the former instance, it is considered to be in the employer's interest, while maintaining the option to determine, for the contractor's employment not to be automatically determined – in the hope that work on site will not stop, that a rescue bid will preserve the company and that this will lead to the most economic completion of the works. In the case of the demise of the company, the automatic determination of the contractor's employment remains.

The business rescue-orientated insolvency procedures include administrative receivership, administration and voluntary arrangements or compositions. The insolvency procedures leading to the inevitable demise of the company include the appointment of a provisional liquidator, compulsory liquidation and a creditor's voluntary liquidation.

JCT98 provisions

Clause 27 provides for the determination of the contractor's employment in the event of default (27.2) and corruption (27.4) as well as in the event of insolvency (27.3). The procedure to be followed by the employer and the contractor is set out in clauses 27.3, 27.5, 27.6 and 27.7.

Clause 27.3 identifies two scenarios. On the one hand, it identifies the various events which lead to the contractor's employment being *'forthwith automatically determined'* (without notice), but with the proviso for reinstatement, if both parties agree – the inevitable demise scenario. On the other hand, it identifies those events of which the contractor has to give written notice and which give the employer the option, upon written notice, of determining the contractor's employment – the business rescue scenario.

Clause 27.5 applies only to the business rescue scenario, that is, where the contractor's employment is not automatically determined and the employer retains the option of determination. The clause relieves the employer from his obligations to make further payments to the contractor and relieves the contractor of his obligation to carry out and complete the works until either an agreement (called a '27.5.2.1 agreement') is reached on the continuation, novation or provisional novation of the contract or the employer determines the employment of the contractor. The clause also makes provision for the employer to ensure that the site, the works and the materials on site are adequately protected.

Clause 27.6 deals with the consequences of determination of the contractor's employment, whether it be automatic or at the employer's option. The clause covers the completion of the works, the making good of defects, the use of temporary buildings and plant, the direct payment of suppliers and sub-contractors and the computation and settlement of the final account.

Clause 27.7 deals with the procedure in the event that the employer decides not to complete the works and is further mentioned in this chapter under the heading 'Final account'.

Early warning signs and precautionary measures

The design team will usually be aware of the contractor's financial difficulties before any formal steps are taken in recognition of insolvency. There may be rumours circulating around the site. There may be a rapid depletion of the contractor's resources, such as materials, temporary buildings or the workforce. Although the contractor may be, or appear to be, in financial difficulties, he need not necessarily be insolvent. In these circumstances the matters should be kept strictly confidential, not only to avoid the possibility of defamation actions but also to prevent the spread of such information, which could quickly drive the contractor into more serious difficulties or into liquidation.

In such circumstances it is particularly important for the architect to check, before the issue of each interim certificate, that the contractor has provided reasonable proof of payment to nominated sub-contractors of the amounts directed as payable to them in previous certificates. When such proof is not forthcoming and the architect is satisfied that it is not because of a failure of the nominated sub-contractors to supply the relevant evidence, he must issue a certificate stating the amount(s) in respect of which the contractor has failed to provide such proof. Provided that the architect has issued this certificate, the employer has a duty, under clause 35.13, to pay the nominated sub-contractors direct and to deduct the amount from any sums due to the contractor. In accordance with clause 35.13.5.3.4 this duty will cease to have effect absolutely if a petition has been presented to the court for the winding up of the contractor, or if there has been a resolution properly passed for the winding up of the contractor, other than for the purposes of amalgamation or reconstruction. In these circumstances there is no obligation on the part of the employer to make any further payment to the contractor (or direct to any nominated sub-contractors) until the notional final account has been prepared.

The first moves ...

Clause 27.3.1 identifies the various events which constitute insolvency under the contract, depending on whether the contractor is an individual or a company.

Clause 27.3.2 covers the giving of written notice to the employer where the contractor has made a composition or arrangement with his creditors or, being a company, has made a proposal for a voluntary arrangement for a composition of debts or scheme of arrangement to be approved in accordance with the Companies Act 1985 (amended by The Companies Act 1989) or the Insolvency Act 1986 – the business rescue scenario.

Clause 27.3.3 covers the automatic determination of the contractor's employment where a provisional liquidator or trustee in bankruptcy is appointed, or a winding-up order is made, or the contractor passes a resolution for voluntary winding up, except for the purposes of amalgamation or reconstruction – the inevitable demise scenario.

Whereas in the former case the employer may, at any time, unless a '27.5.2.1 agreement' has been made, give notice to the contractor to determine his employment under the contract, in the latter case there is no need for any formal notice; this is made clear by the use of the words *'forthwith automatically'* in the contract. An employer should, however, take immediate steps to tell the contractor, as well as his liquidator, provisional liquidator or receiver, that he regards the employment of the contractor as determined in accordance with the relevant clause. This should be sufficient to avoid any suggestion that, by carrying on as if the contractor's employment had not been determined, it has been *'reinstated'*, as referred to in clause 27.6, and which would otherwise be covered by a '27.5.2.1 agreement', referred to more fully later in this chapter.

It is also important to note that it is only the *employment* of the contractor which is determined, not the contract itself. The contract continues to protect the rights of the parties and govern the procedure to be followed.

The principal actions which the design team should take, on behalf of the employer, within the first few days following the insolvency of a contractor are as follows:

- secure the site as far as possible in a cost-effective way to prevent vandalism and unauthorised removal of materials by sub-contractors and suppliers
- list all unfixed materials (whether or not 'site materials'), with a note of those paid for by the employer
- establish the position regarding unpaid certificates, retention, performance bonds and other contractual entitlements
- establish with professional advisers the objectives and determine the employer's preferred route forward
- make arrangements, if appropriate, to allow the contractor, sub-contractors and suppliers access to the site during working hours to continue with the works
- prepare a programme and an estimate of cost for the completion of the works.

Safeguarding the site, the works and materials on site

Ideally as soon as the contractor's employment has been determined, the site should be closed and no materials or plant should be removed, despite the various demands of suppliers and sub-contractors for the return of materials. An understanding of the position regarding ownership of materials is therefore important.

It is a well-established rule that ownership of all materials and fittings, once incorporated in or affixed to a building, passes to the owner of the freehold. Although the employer may not be the freeholder, this should make little difference, in practice, to this rule. Ownership in unfixed materials and goods, however, can be more complex and there is no easy method of determining title. The relevant clauses in JCT98 are 16 (materials and goods unfixed or off site) and 30 (certificates and payments).

The architect has a duty to certify the total value of the materials and goods delivered to or adjacent to the works, provided that they have been properly and not prematurely delivered and are adequately protected (clause 30.2), but he may certify materials or goods before delivery only if they are 'the listed items' (clause 30.3).

Clause 16.1 provides for the property in unfixed materials and goods to pass to the employer, where their value, in accordance with clause 30.2, has been included in a certificate and paid for by the employer.

Clause 19.4 provides that, where the value of a sub-contractor's materials has been included in a certificate and this has been paid, such materials or goods shall become the property of the employer, and the sub-contractor shall not deny this. Provision is also made within JCT98 for the ownership of materials and goods to pass to the contractor where he has paid a sub-contractor in advance of the architect certifying such payment (clause 19.4.2.3 refers – see also clause 4.15.4.3 of NSC/C and clause 21.4.5.3 of DOM/1).

In practice, it is probably safest to assume that all materials on site are in the ownership of the employer and to allow nobody on to the site under any pretext whatsoever. It will then be up to the liquidator or to a sub-contractor or supplier to prove that ownership in particular materials has not become vested in the employer. There is, of course, always the risk that if the title to materials and goods has not passed to the employer, the supplier or sub-contractor may seek an injunction for release of the materials and goods in question or damages for their retention or conversion. If there is the possibility of such a dispute, legal advice should be sought.

If a clerk of works is employed, he may be able to act as a watchman while he is on site. However, it would be prudent for the employer to engage a separate watchman to cover periods when the clerk of works is away from the site. It would also be prudent for the quantity surveyor to make a list of the unfixed materials on the site – he will need this in any event in determining what is necessary to complete the works. The project manager/architect should immediately try to arrange with the employer for all valuable materials which can be moved to be secured under lock and key and for locks to be changed where materials are already secured.

'Listed items' off site

Clause 16.2 provides that, where the value of any 'listed items' intended for the works and stored off site has been certified in accordance with clause 30.3 and paid for by the employer, such 'listed items' shall become the property of the employer. However, the question of title in respect of materials and goods stored off site can be complicated. For example, the contractor would have no title to the materials and goods where these are the subject of a retention of title clause.

Such a clause stems from the judgment in the case of Aluminium Industrie Vaassen BV v Romalpa Aluminium Ltd (1976). In its most basic form, such a clause will state that title to the materials and goods will be retained by the supplier unless and until they have been paid for. The issue may be further complicated where the supplier is divested of title to the goods by virtue of section 25 of the Sale of Goods Act 1979. This section provides, in effect, that the contractor can, by reselling the goods to a bona fide purchaser as a 'buyer in possession', pass a good title, even though the contractor himself lacked a good title at the time. This is a complicated area of the law and it would be wise to seek legal advice in the event that a dispute over title becomes likely.

Completion of the works

It will be necessary to arrange a meeting with all concerned to consider the best method of completing the contract. The employer, his professional advisers and all consultants should be at this meeting. When a liquidator has been appointed he should be kept informed of decisions which have been made. Similarly, if a bond is coupled with the contract, the bond holder should be kept informed. It would be wise, in the light of court decisions, for the employer to check the terms of the bond immediately on becoming aware that the contractor is in financial difficulties. The requirements under the bond, for example as to notice, should be complied with.

There are three options available for completing the works – assuming that the employer wishes to do so and does not choose, following the determination of the contractor's employment, not to complete the works, in which case clause 27.7 would be operated. Assuming completion is intended, the options are:

- reinstatement of the original contractor's employment
- novation – in the form of a '27.5.2.1 Agreement'
- a separate completion contract arranged by the employer.

Reinstatement of the original contractor's employment

Under clause 27.3.3, after the contractor's employment has been automatically determined, the parties can agree to the contractor's employment being reinstated. In these circumstances, the works continue under the original arrangements 'as if nothing had happened'. The contractor is subject to all his original obligations (including completion on time) and the employer maintains all his original rights (including those relating to liquidated and ascertained damages).

Novation – in the form of a '27.5.2.1 Agreement'

Following written notification under clause 27.3.2, the parties enter an interim period under clause 27.5, during which their original obligations are suspended and they are free to make interim arrangements for the work to continue, pending a decision to formalise a new arrangement or terminate the contractor's employment.

Clause 27.5.2.1 makes provision for this new agreement to be a continuation of the original contract, a novation or a conditional novation. This allows sufficient flexibility for the parties to negotiate almost any agreement that suits them and the subsequent contract will be governed by that agreement. It should be remembered that in most of these situations, an insolvency practitioner will be acting on behalf of the contractor.

A separate completion contract arranged by the employer

Following determination of the contractor's employment, whether it be automatic or by the employer's option, the provisions of clause 27.6 become operative, allowing the employer to complete the works as economically and expeditiously as possible – assuming that he wishes to do so and does not decide not to do so under clause 27.7.

To facilitate the completion of the works, clause 27.6.1 provides for the employer to employ and pay other persons to carry out and complete the works. Such persons may use all temporary buildings, plant, tools, equipment and materials and purchase all additional materials necessary to complete the works and make good defects. These rights of the employer, when based upon a determination through insolvency, are dependent upon, first, a valid right in law to use such buildings, plant, tools, etc., and second, the owners of the various items being bound by the contract. In the first case, it is often argued that this clause is void against the liquidator who, as a matter of law, upon his appointment takes into his custody and control all the property of the company. In the second case, many items of plant are hired by the contractor or owned by a subsidiary company which is not a party to the contract. In this latter case it will be necessary to enter into a new agreement with the owners of the hired equipment.

These factors will be taken into account in putting together the documentation and agreeing the arrangements with the completion contractor. Much will depend on the extent of the works to be completed:

- If the works have not started, it will be necessary to appoint a contractor and to fix a new price for the works. This can be done by negotiating with the second lowest tenderer or, if that fails, by seeking new tenders, preferably from those contractors on the original tender list. In this instance there should be no need to amend the original tender documents other than to revise the dates for possession and completion.
- If the works have started but are not complete, it will be necessary to appoint a contractor and to fix a price for the completion works. This can be done either by negotiating with a single contractor or by seeking competitive tenders. In either instance the original tender documents can be used but they will need amending so as to reflect the extent of the works properly completed by the original contractor prior to determination, the materials, goods, plant and equipment available on site for use in the completion works, the method for dealing with the making good of defects arising from the original contract (see later in this chapter), the new dates for possession and completion and any appropriate revision to the amount of liquidated and ascertained damages.
- If the works are substantially complete or complete apart from the making good of defects, it will still be necessary to appoint a contractor for the

completion works. In this situation the remaining work is likely to be of a jobbing nature, which is perhaps best done under a prime cost contract with a fixed fee or, if available, by the employer's own direct labour organisation or maintenance department. In either instance completely new tender documents will be required.

In any event the design team will need to carefully assess the work to be done, including that of sub-contractors, before making recommendations as to the most suitable procedure and contract for completion works. The employer should ensure that the completion works are carried out in the most economical way if disputes with the bond holder or liquidator are to be avoided. He must also ensure that, for the purposes of CDM94, a principal contractor is always in post while any work is in progress on the site (see Chapter 1).

Completion documentation

One item that is essential in any scheme for completion of the work, except where the contractor's employment is reinstated, is a method for dealing with the making good of defects which are bound to arise from the original contract. There are three options available:

- a lump sum premium
- a provisional sum
- measured quantities for known defects, plus a lump sum premium or provisional sum for other items.

Lump sum premium

In certain cases, the most satisfactory completion document is one on identical terms to the original contract, with the addition of a fixed lump sum premium to cover the new contractor's costs in taking over the partly completed works, including price fluctuations, the responsibility for making good defects and a credit for any materials, goods, plant and equipment on site which belong to the employer. This lump sum premium is then paid to the contractor by instalments as the work proceeds. This method has the advantage of identifying the employer's total commitment at the start and avoids the need to account for the cost of remedial works and therefore saves professional fees. Unfortunately, this method cannot be used if the new contractor is not able to make a proper assessment of the risks involved – usually where the remedial works cannot be defined.

Provisional sum

Where the remedial works cannot be defined, a provisional sum has to be included to cover the cost of any such work which comes to light during the course of the completion contract. JCT98 requires the architect to issue instructions in regard to the expenditure of provisional sums. Variations arising from such instructions fall to be valued according to the normal rules for valuation of variations. This system has the disadvantage of an unknown commitment by the employer, but avoids the contractor having to assess, and therefore price, the risk of carrying out the remedial works.

Measured quantities for known defects, plus a lump sum premium or provisional sum for other items

When certain defects are known to involve extensive remedial work it is good practice to incorporate details of the necessary remedial work in the tender documents for the completion works. The cost of making good the remaining defects can then be incorporated either by way of a lump sum premium or a provisional sum. This is a 'halfway house' solution, which has the advantage of identifying a large element of the remedial works costs and reducing the element left at risk to either the contractor (in the case of a premium) or the employer (in the case of the provisional sum). It may well be an appropriate compromise in an uncertain situation, when time available to reach an agreed basis for completing the contract is limited.

Nominated sub-contractors

Under clause 7.10 of NSC/C, if the employment of the contractor is determined under clause 27 of the main contract, the employment of the nominated sub-contractor is also determined. When this happens there are bound to be a number of nominated sub-contractors to deal with and they will fall into two groups:

- Where the works have not been started they will be invited to enter into a similar sub-contract with the new contractor when the new contract is placed.
- Where the works have been partially completed the quantity surveyor should negotiate a price for completing the work, taking care to ensure that only the work to be completed is measured, assessed and valued, making no allowance for the possibility that the nominated sub-contractor may not have been paid for the work already done.

Clearly, it will be in the employer's interest for the sub-contract works to be carried out by the original nominated sub-contractors, but it may be that they will not agree to complete unless they are paid, in full, for all the work carried out under their original sub-contracts. In such instances the design team will have to assess whether or not it would be more cost effective to have work completed by others.

Bond

The contractor may be required to provide a bond (through a bank or insurance company) for the due completion of the work and this is usually set at 10% of the contract sum. Subject to its terms, such a bond was traditionally thought to be available to be used where the employer was put to additional expense as a result of the contractor's liquidation. However, this is now by no means certain in view of findings by the courts that the insolvency of the contractor does not amount to a breach of contract, as there are procedures dealing with it within the contract. Until the uncertainty is removed, this is obviously an area where legal advice should be sought by or on behalf of the employer.

Although the surety may be liable for paying the employer such extra money as may be necessary to complete the job, they have no control over the way the contract is completed. However, the employer should complete the work without undue extravagance and it would be prudent to keep the surety informed as the work proceeds.

Final account

Clause 27.6 deals with the respective rights, duties and responsibilities of the employer and the contractor, so long as the latter's employment has not been reinstated and continued. While this could mean that the contractor would be entitled to payment, in practice this is seldom the case. The employer is entitled to the additional expenses of completing the works, and to any damages which he has suffered by reason of the determination. It appears that this would include any extra fees and expenses which are incurred, and the costs of delay. For example, the architect may well have to make additional visits to the site, and additional prints may be required.

The quantity surveyor will be involved in a great deal of extra measurement, as well as having to sort out complex final accounts. The contractor is responsible for defraying these expenses properly incurred by the employer and the amount of any direct loss and/or expense caused to the employer by the determination, subject to receiving a credit for what he would have been paid had his employment under the contract not been determined. Thus, two final accounts have to be prepared:

- The first, or notional, final account will be the amount that the contract would have cost had the first contractor completed in the ordinary way. This should include variations ordered during both the original and the completion contract, all priced at the rates relevant to the original contract.
- The second final account will be a normal final account for the completion contract, but will include only those variations ordered during the completion contract. These will be priced by reference to the pricing structure of the completion contract, the rates of which are likely to be different from those in the original contract.

Care should be taken to ensure that if the completion contractor has to make good any defects arising from the original contract, the cost of such works, priced as a premium or treated as variations on the completion contract, does not appear as a variation in the first final account; this is because they would not have been so included if the first contractor had completed normally.

The agreement of the second final account should not create any difficulties because there is a contractor's organisation with which to work. But the first final account will almost certainly have been prepared by the quantity surveyor without the assistance of the first contractor, whose staff, following liquidation, will probably no longer be available. The liquidator should therefore be kept informed of the method used to prepare the account and should be sent a copy on completion for his information.

Two examples of a statement showing the financial position of the parties at completion are given at the end of this chapter. In Example 18/1, it has been assumed that the contract was completed without too much trouble and that there is still something to be paid to the contractor in liquidation. It has also been assumed that the extra cost of completing the contract was less than the monies outstanding to the contractor at the time of the liquidation and therefore there is finally a debt due from the employer to the liquidator. In Example 18/2, it has been assumed that, with the contract only one third complete, the extra cost of completing the work was considerably greater than the money outstanding at the time of the liquidation. Almost certainly, the employer suffers a loss; how much of this he can recover depends on the amount in the pound paid by the liquidator and any amount contributed by the bond holder, if applicable.

Clause 27.7 covers the procedure in the event that the employer decides not to complete the works. In such circumstances, he is required to notify the contractor within a period of six months and thereafter, within a reasonable time, to send a statement of account to the contractor.

The procedure upon the insolvency of a nominated sub-contractor

Clause 35.24.7 states that where a nominated sub-contractor becomes insolvent and his employment is determined, the architect shall make such further nomination of a sub-contractor, in accordance with clause 35, as may be necessary to supply and fix materials or goods or to execute the work and to make good or to re-supply or re-execute any defective materials or defective work.

The delay inevitably occurring due to the departure (or dropping out) of a nominated sub-contractor does not entitle the main contractor to any extension of time under clause 25.4.7. However, clause 35.24.10 requires the architect to make the further nomination of a sub-contractor within a reasonable time of the obligation having arisen. Any period of delay attributable to a failure of the architect to nominate in a reasonable time would be dealt with under clause 25.4.6.2. Therefore, an architect has power to extend the contractor's time for completion if a nominated sub-contractor goes into liquidation and there is subsequent undue delay in re-nomination. Also, the contractor would be entitled to reimbursement of any direct loss and/or expense incurred by virtue of clause 26.2.1.2.

The procedure upon the insolvency of the employer

Clause 28.3 deals with the insolvency of the employer in a manner similar to the way that clause 27.3 deals with the insolvency of the contractor.

Clause 28.3.2 requires the employer to inform the contractor in writing if he makes a composition or enters into a voluntary arrangement.

Clause 28.3.3 provides for the contractor to give notice to the employer determining his employment following any act of insolvency, bearing in mind that, notwithstanding any formal notices following any of the events set out in clause 28.3.1, the obligation on the contractor to carry out and complete the works in compliance with clause 2.1 is suspended.

Clause 28.4 covers the consequences arising from determination for any reason, including the employer's insolvency. In essence, the contractor is required to remove all his temporary buildings, plant, tools, equipment and site materials safely, and to ensure that sub-contractors do likewise.

As far as the works are concerned, they are unlikely to proceed, and clause 28.4.3 requires the contractor to prepare an account for submission to the employer and, after taking into account amounts previously paid under the contract, the employer shall pay to the contractor the amount properly due within 28 days of submission.

The employment of the design team will not necessarily end with the employer's liquidation – their terms of appointment should be checked. However, if they are to continue, they should seek undertakings from the liquidator that they will be paid for their services. If the liquidator is prepared to give such undertakings they can then proceed with their respective duties in the winding down of the contract, the settlement of accounts and the resolution of claims.

	£	£
Amount of original contract (with company in liquidation)		200,000
Additions (whether ordered with company in liquidation or completion contractor - priced at rates in the original contract)		10,000
		210,000
Omissions (whether ordered with company in liquidation or completion contractor - priced at rates in the original contract)		7,500
Amount of notional final account if original contractor had completed		202,500
Amount of completion contract	30,000	
Additions (for completion contract only - priced at rates in the completion contract)	2,500	
	32,500	
Omissions (for completion contract only - priced at rates in the completion contract)	2,000	
Amount of final account of completion contract	30,500	
Additional professional fees incurred	1,500	
Amount certified and paid to original contractor before liquidation	169,000	201,000
Debt payable by employer to liquidator or receiver		£1,500

Example 18/1 Financial statement at completion showing a debt from employer to contractor.

	£	£
Amount of original contract (with company in liquidation)		200,000
Additions (whether ordered with company in liquidation or completion contractor - priced at rates in the original contract)		10,000
		210,000
Omissions (whether ordered with company in liquidation or completion contractor - priced at rates in the original contract)		7,500
Amount of notional final account if original contractor had completed		202,500
Amount of completion contract	150,000	
Additions (for completion contract only - priced at rates in the completion contract)	10,500	
	160,500	
Omissions (for completion contract only - priced at rates in the completion contract)	7,700	
Amount of final account of completion contract	152,800	
Additional professional fees incurred	7,500	
Amount certified and paid to original contractor before liquidation	72,500	232,800
Debt payable by liquidator or receiver and partly by bond holder (if applicable) to employer		£30,300

Example 18/2 Financial statement at completion showing a debt from contractor to employer.

Index